THE
SEARCH
for the
VIRUS

Steve Connor and
Sharon Kingman

PENGUIN BOOKS

We stand nakedly in front of a pandemic as mortal as any pandemic there has ever been.

> Dr Halfdan Mahler, Director-General of the World Health Organization, speaking on AIDS at the United Nations on 20 November 1986

(*Pandemic*: an extensive epidemic, covering a wide geographical area.)

PENGUIN BOOKS

Published by the Penguin Group
27 Wrights Lane, London W8 5TZ, England
Viking Penguin Inc., 40 West 23rd Street, New York, New York 10010, USA
Penguin Books Australia Ltd, Ringwood, Victoria, Australia
Penguin Books Canada Ltd, 2801 John Street, Markham, Ontario, Canada L3R 1B4
Penguin Books (NZ) Ltd, 182–190 Wairau Road, Auckland 10, New Zealand

Penguin Books Ltd, Registered Offices: Harmondsworth, Middlesex, England

First published 1988

Filmset in Linotron 202 Melior

Typeset, printed and bound in Great Britain by
Hazell Watson & Viney Limited
Member of BPCC plc
Aylesbury Bucks

PENGUIN BOOKS

THE SEARCH FOR THE VIRUS

Steve Connor has a degree in zoology from the University of Oxford. He joined *New Scientist* as deputy news editor in 1983. He is co-author of a book on surveillance, computers and privacy entitled *On the Record* (Michael Joseph, 1986). In 1986 he won a Science Writers Fellowship.

Sharon Kingman has a degree in zoology from the University of Durham. Since then she has spent ten years in medical and scientific journalism. She joined the features department of *New Scientist* in 1985, becoming science correspondent at the end of 1986.

CONTENTS

LIST OF FIGURES

Illustrations by Peter Gardiner

ACKNOWLEDGEMENTS

We thank the many people who have helped us to write this book. In particular, we express our gratitude to Fred Pearce, who read and made useful comments on several chapters; to Celia Dodd, who gave expert editorial advice as a non-scientist; and to Susanna Hourani and Peter Gardiner,who made helpful suggestions about parts of our early draft. We also thank the many scientists at the forefront of AIDS research who read and criticized certain chapters of the book, or took part in valuable discussions, including Michael Adler, Mike Bailey, John Barbara, Wilson Carswell, Angus Dalgleish, Anne Johnson, Timothy Peto, Quentin Sattentau and Ian Weller. All opinions and errors, of course, remain our own. Caroline Akehurst, David Fitzsimons and Renee Sabatier provided useful assistance. Omar Sattaur carried out much of the early groundwork on the subject. Christopher Joyce and Ian Anderson provided important research material from the United States. Peter Gardiner produced the illustrations from minimal instructions. Michael Kenward encouraged us to start the project in the first place, as well as commenting on our manuscript. Finally, we thank our colleagues on *New Scientist* who have tolerated our preoccupation with AIDS.

CHAPTER 1

AN INFECTION OF IGNORANCE

We could not have designed a more frightening disease if we had tried. If we could play at being Satan for the day, charged with the task of designing an epidemic to undermine both the developed and underdeveloped countries of the world at the end of the twentieth century, then the blueprint for the design would incorporate many of the features of AIDS. An artificial pestilence would have to have all the subtleties and idiosyncrasies of acquired immune deficiency syndrome – AIDS – in order to kill the maximum number of people and cause untold disruption to the health services of countries throughout the world.

A modern plague has a greater chance of spreading if it can be transmitted during the most intimate and compulsive of human activities – sex. Science can fight disease with the cooperation of those at risk of infection, but a disease that spreads with the help of sex is a formidable foe. There are few occasions in history when laws have successfully stopped people from having sexual relations with each other. And in modern society, sex has gone international. People travel more now than ever before, and sexual relationships have crossed national boundaries. Sex has even become a tourist attraction. Yet, despite the sexual liberation that has marked the latter half of the twentieth century, sex is still a surreptitious activity, so the hypothetical designers of a modern

1

plague can take advantage of the age-old taboos that are still associated with sex. The shame of sex and the reluctance even to admit its existence is, after all, as ancient as the story of forbidden fruit in the Garden of Eden.

A modern plague would have to be more subtle and complex than previous diseases in order to outwit modern medical science. It would have to attack an important and yet vulnerable part of the body. AIDS does this. The virus that causes AIDS attacks and destroys the immune system, which is responsible for fighting off infections. Without the immune system, the human body cannot fend off even the most trivial illnesses, and it becomes an open target for hundreds of potentially deadly organisms. AIDS results from the inability of the body to defend itself from fatal infections. A person without an effective immune system becomes easy prey to a myriad of microorganisms, which can confuse doctors and scientists trying to make an accurate diagnosis of an underlying cause.

The structure of the AIDS virus is almost perfect. For a start, in common with all viruses, it is tiny; so small that its simple structure cannot be said to be living because it can only replicate by hijacking a living cell. A virus therefore straddles the divide between living and non-living matter. The virus lives and replicates within human cells, a perfect place to escape from the body's defences. Most viruses do this, which is why they are so effective. The virus that causes AIDS, however, performs an even neater trick: it becomes part of the genetic material of the cell. Hiding within this innermost sanctum of our bodies, the virus has adopted the ultimate camouflage – it becomes part of the person it infects.

Cloistered away like this, the virus becomes quiescent. Its dormancy can last months or even years. People who carry the virus during this time have few, if any, obvious symptoms of infection. Some may never go on to develop the hotch-potch collection of diseases, such as skin cancer, pneumonia, and other infections, that comprise acquired immune deficiency syndrome. (These terminal infections signal 'full-blown' AIDS.) People without the early signs of AIDS can

2

nevertheless still pass the virus to others. With no symptoms to alert them, carriers can be completely unaware that they are spawning further generations of viruses. A designer of a modern plague would admire this characteristic of AIDS too. The long delay between the moment of infection and the appearance of the symptoms of AIDS secures the stealthy spread of the virus throughout the population.

Another strength of the virus's design is its ability to change its structure with each generation. The body's immune system relies heavily on being able to recognize microorganisms from their outer coatings. The AIDS virus mutates very fast, so fast that it becomes exceedingly difficult to identify any similarities between the outer coat of one individual virus and the coat of another. The result is that the body cannot launch a successful attack against this constantly moving target. This detail completes the frustratingly perfect design of the virus that causes AIDS.

The perfection, however, does not end with the virus, or with how it spreads from person to person. Other aspects of AIDS make it difficult for modern science to tackle. There are further prejudices and taboos, not just about sex, but about sexual preferences, lifestyles, drug addiction, race and nationality. These prejudices combine to make AIDS the ultimate vehicle for transporting fear, hysteria, hatred and panic around the four corners of the globe. If scientists could not have designed a better disease, then psychologists could not have designed a better mechanism for provoking irrational reactions and disturbing the human psyche.

The story of AIDS is full of examples of ignorance. This ignorance has led to confusion and fear, which in turn have developed into distrust and hatred. AIDS does not spread by casual contact. People cannot become infected with the virus by handling the clothing of people with the disease or by shaking hands with them. AIDS is not, in short, a highly contagious disease, so it is not strictly speaking a 'modern

3

plague', equivalent to the Black Death which decimated Europe in the Middle Ages.

Medical scientists became concerned about the early press reports on AIDS, which often described the disease as the 'gay plague'. The press coverage seemed to be encouraging irrational fears: people were being led to believe that they could catch AIDS by merely associating with sufferers. As a dozen of the world's leading figures in medical virology and infectious diseases made clear at a meeting at the World Health Organization in Geneva in September 1985, one of the greatest threats of AIDS arises from ignorance, in particular ignorance about how the disease spreads. The chairman of the meeting, Professor Friedrich Deinhardt, a leading virologist from the Max von Pettenkofer Institute in Munich, said at the time: 'In no way can AIDS be compared to the great plague of the Middle Ages. There is no evidence that it is spread through casual contact with an infected person. It is primarily a sexually transmitted disease.'

Scientists wanted to nip the growing hysteria about AIDS in the bud. It would take a long time for the message to seep in. Part of the problem was that far too many people in a position to know better were still confused about how AIDS can spread from one person to another.

Such ignorance fuelled the existing prejudice against homosexuals and other groups associated with AIDS. Signs of AIDS hysteria range from the serious to the trivial. In 1985 telephone engineers initially refused to mend the equipment at Gay Switchboard in London. The engineers said that they might catch AIDS from the telephones.

Haemophiliacs, who suffer from a clotting disorder of the blood and who also became a risk group for AIDS, did not escape. Haemophiliac children were ostracized by their school friends. Worse still, some parents of healthy children tried to stop haemophiliacs from attending the same school as their children. In Indiana in the US, for instance, pressure from parents prevented a nine-year-old haemophiliac boy from attending lessons. In a similar case in New York, parents kept thousands of children at home as a protest at the local school authorities permitting children with AIDS to attend classes. In Britain, and other European countries, similar concern grew about the perceived danger that haemophiliacs posed to other children. In one school in Hampshire in southern England, for instance, many parents kept their children at home for fear of their catching AIDS from a pupil at the school, a small boy with haemophilia.

The British government came under pressure from the worried parents of haemophiliac children. The Department of Health issued a pamphlet to teachers in March 1986 explaining that children with the AIDS virus should be treated no differently from other children, and could take part in school activities in the usual way. This leaflet explained that there is a danger of children passing on the virus in blood, so teachers should discourage any behaviour that may result in bleeding, such as tattooing, ear piercing or 'blood-brother' rituals. In many cases, however, the advice did little to quell the growing fear of other children and their parents.

The fear of AIDS has resulted in hysterical reactions in supposedly well-educated and well-informed groups of people. For example, the dons at one Oxford college banned a centuries-old tradition of passing a 'loving cup' of wine around table at the college's annual dinner. They thought the ban would be prudent given that 'certain diseases' were rife, even though there was no scientific evidence that the virus could spread in saliva. In the US, the National Science Foundation announced in 1987 that scientists wanting to spend the winter at its research post in the Antarctic must have a blood test for infection with the AIDS virus. The regulation did not apply to scientists wanting to spend summer there – it appears that even some scientifically trained people were under the impression that sex only takes place after dark.

Accusations of overreaction have been levelled at medical authorities. For example, in 1987 some Scottish medical schools advised students not to go to African countries for fear of infection with the AIDS virus. Several doctors from the same medical schools pointed out in letters to the British medical journal, the Lancet, that the risk of acquiring AIDS

5

as a result of emergency treatment with unsterilized surgical instruments or contaminated blood is more remote than the usual risks associated with travelling in tropical climates. Two of these doctors, Arnold Klopper and Nicholas Fisk of the University of Aberdeen, wrote: 'It is a remarkable coincidence that such controversial advice should have been issued simultaneously by three or four Scottish medical schools . . . To ask us to ostracize our colleagues in Africa is a serious matter. It marks a profound departure in university attitudes and policy.'

African countries have borne the brunt of a rising tide of xenophobia resulting from the spread of AIDS. Many governments outside Africa have stipulated that foreigners must have a test for infection with the AIDS virus before they can enter and stay in the country in question. Very often the testing regimes were ill-disguised attempts to take unnecessary and arbitrary action against visitors and students from African countries. India, for example, initially wanted to test African students resident in the country. The Indian government later amended this to include all foreigners, except diplomats and journalists, because of the blatantly discriminatory nature of the initial plan.

Throughout the world, governments have moved to stem the spread of AIDS. A common thread running through the laws that they have enacted is the belief that testing foreigners will somehow limit the incidence of AIDS at home. In Britain the government has so far resisted a vociferous campaign to test black immigrants. The campaign reached a shrill climax with the leaking of a 'Whitehall report' to the *Sunday Telegraph* in September 1986. The report was evidently a collection of impressions of British diplomats stationed in several countries in central Africa. The author of the report, which went to the Foreign Secretary, suggested that visitors from Africa 'could be a primary source of infection and should be subject to compulsory tests'. An editorial in the *Sunday Telegraph* articulated the paper's position:

'It would be monstrous if ministers were to hold back from action lest they be accused of racial discrimination. For in this instance, there is a positive duty to discriminate. In the matter of AIDS, black Africa does have a uniquely bad record and only harm can spring from pretending otherwise.' (The editorial invoked the familiar theme of invading black hordes. It said: 'In a few days' time, hundreds of students from Zambia, Uganda and Tanzania will be arriving in this country. A significant proportion of them – possibly up to 10 per cent – could be AIDS carriers.')

The real pretence, however, was that countries could somehow close the stable doors after the horse had bolted. Britain by this time had already become seriously infected with the virus, and the vast majority of infected people had had no contact whatsoever with 'black Africa'.

In the US, the country with more documented cases of AIDS than any other, the same irrational view came to the fore in May 1987 with the first speech on AIDS by President Ronald Reagan. Before a glittering dinner held at the Potomac Hotel in Washington DC to raise funds for the American Foundation for AIDS Research, chaired by the actress Elizabeth Taylor, Reagan upset many scientists by saying that he supported routine testing for infection with the AIDS virus. The medical establishment had already reached a virtual consensus that routine testing was impractical, expensive and next to useless. Yet Reagan said:

'I have asked the Department of Health and Human Services [HHS] to determine as soon as possible the extent to which the AIDS virus has penetrated our society and to predict its future dimensions. I have also asked HHS to add the AIDS virus to the list of contagious diseases for which immigrants and aliens seeking permanent residence in the United States can be denied entry.' By the following month, June 1987, the US had classified AIDS as a dangerous 'contagious' disease under the Immigration and Nationality Act. Anyone suffering from AIDS would be denied permanent entry into the US. Furthermore, the US government also decided to consider classifying people infected with the virus, but without AIDS, in the same category. Many scientists, however, were still not convinced of the usefulness of

7

blood tests as a technique for mass screening. A negative test, for instance, is not proof that the person is not infected. And when screening a large group of people for a virus that is very rare, there is a strong chance of wrongly identifying uninfected people as 'positive'.

Countries with diverse political complexions have taken very similar stands on testing foreign visitors. South Africa, for example, announced in 1987 that all immigrants, including hundreds of thousands of black migrant workers, would have to have tests. Those found positive would be expelled. Cuba has drawn up similar plans to test foreign visitors. The Cuban Deputy Minister of Health, Hector Terry, said in 1987 that all of Cuba's ten million citizens will be tested by 1989. Terry has also issued warnings to Cubans to 'avoid fortuitous contacts with foreigners', whatever that may mean.

In China, fear of AIDS and of foreigners resulted in a government campaign to burn imported second-hand clothes. In November 1985, a crowd of city officials from Beijing, including the city's deputy mayor, watched a group of soldiers with flamethrowers incinerate twenty tonnes of second-hand clothes imported from Japan, Hong Kong and Macao. The officials feared that the clothes harboured the AIDS virus.

In the Soviet Union, the Presidium of the Supreme Soviet, the highest authority in the country, took the necessary legislative steps to ensure that rigorous measures could be taken against anyone, foreign or native, infected with the AIDS virus. The Presidium adopted a decree in August 1987 stating:

The citizens of the USSR, as well as foreign citizens and stateless persons living or staying in the territory of the USSR, may be bound to take a medical test for the AIDS virus. If they dodge the test voluntarily, the persons, in relation of whom there are grounds for assuming that they are infected with the AIDS virus, may be brought to medical institutions by health authorities with the assistance, in the necessary cases, of authorities from the interior ministry.

The decree stated that anyone infected with the virus who is found to have deliberately exposed other people to the virus would be imprisoned for up to five years. For actually giving AIDS to someone else knowingly, the maximum prison sentence is eight years.

Soon after the government issued this decree, the Soviet authorities put the new law into action. In September 1987, more than a hundred foreign visitors were expelled from the country. All but three were from central Africa. In the same month, a twenty-eight-year-old Soviet woman had to make a pledge to refrain from sexual relations for an arbitrary period of five years because she was diagnosed as being infected with the AIDS virus. If she broke the pledge, her punishment would be a prison sentence of up to eight years.

Many countries in the West have also discussed what to do about people who are infected with the virus and who knowingly put others at risk. In West Germany, courts in the state of Bavaria jailed a prostitute in 1987 for continuing her trade when she knew she had the AIDS virus. In the same state, an American civilian was accused of grievous bodily harm for having sexual intercourse even though he knew that he was infected. In November 1987, the courts sent the man, an American citizen, to prison for two years. In Switzerland, doctors are obliged by law to tell health authorities about all cases of infection, not just cases of AIDS. The rule, which became law in 1987, marked a new departure for countries in western Europe, some of which had shied away from making it obligatory to notify even cases of AIDS because of worries about the confidentiality of the information.

Ever since the problem of AIDS first arose, Britain has tussled with the issue of whether to make AIDS a 'notifiable' disease. This would make it compulsory for doctors to notify local authorities of patients with AIDS, and could also give hospitals the right to apply for a court order to detain patients against their will if necessary. In early 1985, the government considered making AIDS notifiable but, after taking advice from his committee of experts, the Minister of Health at the

9

time, Kenneth Clarke, did not think this was necessary. However, Clarke said in February 1985 that:

> There might be very rare and exceptional cases where the nature of a patient's condition would place him in a dangerously infectious state which would make it desirable to admit him to, or detain him in, hospital . . . It is my intention therefore to lay regulations under the Public Health (Control of Disease) Act 1984 which would give reserve powers to authorities to detain a patient when he is in a dangerously infectious condition . . . We need these reserve powers for the very rare case that might eventually arise somewhere sometime.

Such a case arose for the first time seven months later, in September 1985. A twenty-nine-year-old man suffering from AIDS in a Manchester hospital became the first person in Britain to be detained against his will because he had AIDS. The hospital authorities had applied for a court order because the patient was 'bleeding copiously and trying to discharge himself' from hospital. The court order had the backing of Britain's chief medical officer, Donald Acheson, who felt that 'in the circumstances it would be too risky for [the patient] to leave the hospital'. The court eventually lifted the order after an appeal and a promise by the patient to stay in hospital and to continue treatment.

In the US, a vociferous campaign to incarcerate AIDS sufferers gathered momentum after doctors found that the disease had begun to spread to groups outside the gay community. At least one American senator was heard to say that 'somewhere along the line we are going to have to quarantine'. In November 1986, Californians voted on a new law, dubbed 'Proposition 64', that would, if enacted, make it compulsory for doctors to report people carrying the virus or suffering from the disease. The supporters of the proposition, the Prevent AIDS Now Initiative Committee (who did not worry about being called by their acronym, PANIC) wanted no carrier of the virus to be a teacher, employee, or student of a university or school in the state. They also called on authorities to 'quarantine [people] as much as required to stop the spread of the disease'. Compulsory reporting would be necessary in order to identify the people to lock up.

A powerful supporter of Proposition 64 was Lyndon La Rouche, a right-wing political figure in the United States. He justified the proposal with a number of 'facts'. Medical research, LaRouche said, had proved that biting insects spread AIDS, and that people can pass on AIDS by 'casual contact'. Medical research had proved nothing of the sort, and when Californians voted on Proposition 64 on 4 November 1986, they threw it out. This came as a ray of hope for those who feared that ignorance was beginning to dominate the battle against AIDS. After all, only five years earlier, science had been on the brink of overcoming one of the main obstacles to fighting AIDS – ignorance of its existence.

CHAPTER 2

DIAGNOSIS OF A DISEASE

The title of the first scientific paper on AIDS gave little away: "Pneumocystis pneumonia – Los Angeles". The article was published on 5 June 1981, in a then relatively obscure American journal called *Morbidity and Mortality Weekly Report* (*MMWR*), and did not mention the word AIDS. Within a year of that article being published, doctors in the US began to realize that an epidemic of mysterious illnesses had appeared in young, mainly homosexual, men.

Michael Gottlieb and Wayne Shandera knew that something strange was happening to some of their patients. Gottlieb, a doctor with the University of California at Los Angeles, had come across four cases of an exceedingly rare type of pneumonia, which was caused by an infection with a microorganism called *Pneumocystis carinii* in the lungs. All four patients were young men in their late twenties or early thirties, and all were homosexual. He spoke to his colleague Wayne Shandera, of the Los Angeles County Department of Public Health, about the unusual outbreak. Shandera looked at his own records and found that his department had a similar case. This patient was homosexual as well. So there were five cases of *Pneumocystis* pneumonia in the same area, at the same time, and in a group of men who were all young homosexuals. Gottlieb and Shandera wondered if they had discovered a new epidemic.

A vapid account of their work appeared in *MMWR*. 'The patients did not know each other and had no known common contacts or knowledge of sexual partners who had had similar illnesses.' An editorial note at the end of this short report said that the occurrence of *Pneumocystis* pneumonia in the US was almost exclusively limited to patients who had undergone some sort of therapy which had, as a side effect, severely depleted the number of white blood cells of their immune systems. Doctors call this immune suppression or immunodeficiency. A suppressed immune system is seriously weakened to such an extent that infections can take hold of the body. The editorial continued: 'The occurrence of pneumocystosis in these five previously healthy individuals without a clinically apparent underlying immunodeficiency is unusual.'

A month later, on 4 July 1981, another report appeared in the *MMWR* the journal of the organization responsible for monitoring the health of Americans, the Centers for Disease Control, in Atlanta, Georgia. The CDC had gathered together a number of intriguing reports from doctors on both the west and east coasts of America. These doctors had found a very rare type of skin cancer, called Kaposi's sarcoma, in twenty-six homosexual men: twenty from New York City and six from California.

What made the Centers for Disease Control interested in these reports was the age of the men concerned. They were all much younger than the patients who normally develop Kaposi's sarcoma. (This type of skin cancer usually affects men in their seventies, and then tends to develop only in men of certain racial groups, mainly of Mediterranean ancestry.) Seven of the patients had infections as well as the skin cancer – four had *Pneumocystis* pneumonia, just like the five homosexual men that Gottlieb had reported a month earlier. Furthermore, since the publication of Gottlieb's account, doctors in California had found an additional ten cases of *Pneumocystis* pneumonia in homosexual men.

An editorial in the *Morbidity and Mortality Weekly Report* pointed to the extraordinary nature of the observations: 'The occurrence of this number of Kaposi's sarcoma cases during a thirty-month period among young homosexual men is considered highly unusual. No previous association between Kaposi's sarcoma and sexual preference has been reported.'

The CDC decided to set up a task-force to investigate these strange illnesses. One avenue of inquiry was to search through the CDC's files for cases where doctors had prescribed a drug called pentamidine. This drug is often effective against *Pneumocystis* pneumonia. The CDC found that past prescriptions for pentamidine, which the CDC collates centrally for the whole of the US, revealed the existence of more cases of this form of pneumonia among young men. But few cases had occurred before 1979, indicating that the 'epidemic' was a very recent one.

Further work by the CDC, and published in August 1981 in the *MMWR*, confirmed a link between *Pneumocystis* pneumonia and Kaposi's sarcoma: 'The apparent clustering of both *Pneumocystis carinii* pneumonia and Kaposi's sarcoma among homosexual men suggests a common underlying factor.' A year later, in 1982, the CDC came to the conclusion that something was affecting the immune systems of these patients, so permitting 'opportunistic infections'. These are infections which the body can normally shrug off but which take advantage of a weak immune system. An early name for the condition was gay related immune deficiency syndrome—GRIDS. Later, in 1982, the CDC changed this to acquired immune deficiency syndrome, or AIDS, because people other than homosexual men also began to develop the condition. The doctors thought 'AIDS' suitable because people acquired the condition rather than inherited it, because it results in a deficiency within the immune system, and because it is a syndrome, with a number of manifestations, rather than a single disease.

Throughout 1982, and subsequent years, the researchers at the CDC continued to collect details on more and more cases of AIDS. The syndrome was affecting women as well as men, haemophiliacs as well as homosexuals, and children as well as adults. A crisis was unfolding.

The World Health Organization estimated that the virus responsible for causing AIDS had by 1987 infected between five million and ten million people. By the end of 1987, the WHO had heard of more than 60,000 cases of AIDS in more than a hundred countries. This figure, the organization said, represents 'only a fraction of the total cases' of AIDS – such is the difficulty that some countries have in reporting and diagnosing people suffering from the various illnesses that make up acquired immune deficiency syndrome.

The 'hidden pandemic' of AIDS has forced the WHO to establish a special programme to tackle the disease, especially in underdeveloped countries of the Third World. The organization believes that by 1991, one million people on the globe will suffer from AIDS, placing an extraordinary burden on the health services of the developed world, and bringing about the economic collapse of certain countries in the Third World. The director of the WHO's Special Programme on AIDS, Jonathan Mann, spoke about the problem at a scientific conference in Washington DC in 1987. He said that an analysis of the numbers of people estimated to be infected with the AIDS virus and a look at the numbers of people suffering from the symptoms of AIDS led him to the 'virtually inescapable conclusion that we are imminently facing a precipitous increase in the number of AIDS cases'. He continued: 'If five to ten million persons are currently infected [with the AIDS virus] worldwide, and assuming that 10–30 per cent of these persons will develop AIDS during the next five years, then from 500,000 to three million new AIDS cases will emerge from persons already infected with [the virus]. Compared with the number of AIDS cases reported thus far, a more than ten-fold increase in AIDS cases may be anticipated during the next five years.' That was in 1987.

In his speech, Mann referred to the 'two epidemics' of AIDS – the first being the realization that the virus has infected a large proportion of a given population, and the second being the inevitable consequence of this: people start dying of AIDS. The 'third epidemic', he said, follows the first two. 'This is the epidemic of economic, social and political

15

and cultural reaction and response to . . . AIDS.' AIDS is in many ways different to many of the epidemics that the WHO has had to deal with. Most public-health problems in the past have affected either the very young or the very old. But AIDS affects, as Mann said, 'the most vital segment of the population in terms of social and economic development. The selective involvement of young and middle-aged adults, including business and government cadres and members of social, economic and political élites, leads to potential for economic and political destabilization.' What political system, Mann asked in 1987, could withstand up to 25 per cent of its young adults dying of AIDS?

The United States is the richest nation on earth yet the figures suggest that even the US will find it increasingly difficult to meet the economic as well as the social costs of AIDS. By the end of 1987, almost 50,000 Americans had developed AIDS, and many of them had already died. The US Institute of Medicine predicts that by 1991 this figure will have leapt to 270,000, more than five times as many. The institute believes that in 1991 doctors will diagnose 74,000 new cases of AIDS – equivalent to the population of a small town. In the same year, the institute predicts, over 50,000 Americans will die of AIDS. By this stage, most people will personally know of someone who has developed the disease. By 1991, AIDS will have claimed the lives of 180,000 Americans altogether.

There will be regions in the US where the situation will be much worse. Some cities have already reached a crisis. Since 1984, for example, AIDS has become the biggest single cause of death among men aged between thirty and thirty-nine living in New York City. By 1987, AIDS had also become the number-one killer among women aged between twenty-five and thirty-four in the same city. Other cities in the US and elsewhere in the world will inevitably follow the trend. An editorial in the *Journal of the American Medical Association*, in September 1987, described the prospects for the US:

'The loss of human life is staggering. By 1991, AIDS will be among the top ten leading causes of death in the United

States, far exceeding all other causes of death for people between the ages of twenty-five and forty-four years. Socially, politically and economically, AIDS is an ever-worsening disaster.'

In 1981, the year that scientists first discovered a condition they later called 'AIDS', the US government spent $200,000 on the disease. The government expected to increase this allocation to $790m by 1988, two thirds to be spent on medical research to find a vaccine and a cure, and the other third designated for prevention and health education.

The costs to the US, however, of an ever-rising number of people falling ill with AIDS and needing medical treatment are truly staggering. Accountants in the US government estimated in 1987 that the total cost of medical treatment for AIDS sufferers will reach $8,500m in 1991, which will be 1.4 per cent of the nation's total health bill. In 1985, AIDS soaked up just 0.2 per cent of this national expenditure. The same accountants, from the US General Accounting Office, came up with a more worrying calculation. They estimated the amount of money the American economy will lose as a result of young working people dying prematurely of AIDS and calculated that this sum will come to more than $55,000m, by 1991. Even this staggering figure is an underestimate, as accountants did not include days lost from work due to the variety of illnesses people contract before they develop full-blown AIDS. The accountants also based their calculation on the assumption that just 20–30 per cent of the people infected with the virus will go on to develop AIDS, which many scientists believe is a conservative figure. They believe it could be 50 per cent or more.

A study by American researchers in September 1987 found that it cost annually $20,000 on average to look after an AIDS patient who had to spend short periods each year in an American hospital. The researchers came to this conclusion after surveying about 5,400 hospital patients with AIDS two years previously. The researchers, who published their findings in the *Journal of the American Medical Association*, concluded: 'Our results portend possible catastrophic effects on both government and private payers as health-care costs of AIDS escalate.' The journal said in an editorial in the same issue: 'The tremendous costs of this particular disease are exacerbating the problems of an already flawed health-care system . . . Who, then, is going to pay the costs?'

Even the enormous sums that the US government had earmarked in 1987 for AIDS are not going to be enough, according to the report of the disease by the accountants from the General Accounting Office. The spending plans of the US government for projects to prevent the spread of AIDS, in particular, were 'not adequate'. Many specialists involved in fighting AIDS thought that 'perceived lack of federal leadership' was making matters worse and was 'at least as troublesome as estimated shortfalls in the budget'. The US Institute of Medicine had told the General Accounting Office that many students showed an 'alarming degree of misinformation' about AIDS. 'As late as 1986,' the GAO's report states, 'even students in San Francisco were seriously misinformed about routes of transmission and preventive practices. Specifically, 40 per cent did not know that AIDS is caused by a virus or that the use of a condom during sexual intercourse decreases the risk of transmitting [the virus].'

The US government has shown a startling degree of indifference to the epidemic, even though AIDS had claimed more lives in America than the rest of the world put together. It took six years for President Reagan to make a pronouncement on the disease, and a presidential commission he set up in July 1987 was immediately criticized for being preoccupied with morals rather than with tactics to fight AIDS. Many people believed the commission did not fully represent the high-risk groups, notably gays, most in need of a voice in government. Neither did it have enough scientific expertise. The executive director of the commission was forced to resign just two months after taking office after disagreements emerged within the commission itself. Many scientists and others working on AIDS thought that the commission lacked both expertise and objectivity.

In October 1987, the US government was due to have sent

a leaflet on AIDS through the mail to every American household, which most countries in Europe had already done. The brochure, called 'America Responds to AIDS', failed to go out at the allotted time because of disagreements within the administration on how best to educate the American public about the disease. The presidential commission agreed to postpone the plan. Its chairman, W. Eugene Mayberry, told the American journal *Science*, 'we just felt like we weren't ready to tackle the mailing yet'. By comparison to leaflets produced by other governments, notably in Europe, 'America Responds to AIDS' was hardly an explicit piece of literature. As *Science* said, 'the word "homosexual" does not put in an appearance, even though homosexuals comprise 70 per cent of AIDS cases'. Words such as 'family' appeared more frequently than words such as 'condom'.

The failure of the Reagan administration to coordinate a nationwide education campaign early in the AIDS epidemic stands in stark contrast to the dire warnings issued by the administration's senior advisers. The Surgeon-General of the US, C. Everett Koop, said after the World Health Assembly in Geneva in May 1987: 'There's no doubt about the fact that most of us here who are in public health believe that we are facing the greatest challenge perhaps that public health ever faced.' Even the US Secretary of Health in 1987, Otis Bowen, had to agree. AIDS, he said, 'could well become one of the worst health problems in the history of the world . . . an awesome health problem that could involve millions of people who are going to die as a result'. The US government as a whole, however, seemed strangely oblivious to this prophecy.

Elsewhere in the developed world, the scourge of AIDS had not, by 1987, become anything like as severe as in America. Nevertheless, the signs clearly indicated that when doctors looked hard enough, they discovered that an epidemic was quietly developing. France, with more documented cases of AIDS than any other country apart from the US, reported its two thousandth case by the middle of 1987. But by the end of the year, Brazil had usurped France's position as runner-up to the US in the AIDS league table. The silent epidemic had emerged as a strong threat in South America.

In Britain, health authorities reported the thousandth case of the disease towards the end of 1987. The figures showed that the problem of AIDS in the US still dwarfed that in Britain. Nevertheless, the British government had no cause to cheer. Health economists in Whitehall forecast dire consequences for the National Health Service if the numbers of people suffering from AIDS followed the American trend. Officials within the Department of Health had initially warned the government that the cost of treating an AIDS patient would be about £10,000 a year. This was bad enough. A later analysis of the real costs found that this figure was a serious underestimate. The government heard that the true cost of treating an AIDS patient was more like £20,000 a year and this would increase to over £25,000 a year when new, expensive drugs became available for therapy. The government's advisers were forecasting that the cost of treatment could accelerate almost as swiftly as the numbers that would need therapy.

The threat of hundreds of people dying and of the financial burden this would place on the health service, forced the government to set up a special Cabinet committee in 1986. Lord Whitelaw chaired the committee, which was to coordinate the government's response to the disease. The committee included ministers and advisers from most of the departments of government – Health, Foreign Office, Home Office, Education and Science, Defence and, perhaps the most important department of all, the Treasury. Scientists from the Medical Research Council met the committee regularly to keep Whitelaw and other ministers up to date on scientific developments. The Cabinet committee set several precedents. It was the first to be established to tackle a specific disease, and the first to operate in the glare of publicity (the British government normally keeps even the existence of Cabinet committees a secret).

One of the most worrying aspects about AIDS, which the committee soon realized, was the uncertainty of the predictions made by the government's scientific advisers about the course of the epidemic in Britain. An early attempt to gauge the extent of the AIDS problem, made in 1985 by researchers from the Communicable Disease Surveillance Centre (Britain's equivalent to the Centers for Disease Control in the US), which monitors infectious diseases, predicted that 1,800 people would develop AIDS in 1988. The trouble with this informed guess, however, was that the scientists admitted that they could have got it dramatically wrong. There was a one in twenty chance, they said, that the actual number of new cases in 1988 would be less than 460 or greater than 7,300. To a politician having to make decisions affecting the next few years these margins of error are very worrisome indeed.

Whitelaw and his colleagues wanted to know about the extra financial burden that AIDS would put on the state in the years to come. One organization had already assessed the problem. The Office of Health Economics, a private body funded by the drug industry, had estimated, in 1986, that Britain would have to spend between £20m and £30m on treating AIDS patients by 1988. 'But these figures might substantially misrepresent the eventual cost,' the office said unhelpfully. Medical journals had by then already warned that if the government did not treat the growing epidemic of AIDS seriously, then the numbers of people dying from the disease could equal the deaths resulting from the crash of a fully laden jumbo jet each month. The Office of Health Economics suggested another comparison if the government continued its 'crisis-management approach'. The office warned: 'Retrospective judgement might then suggest that an annual loss of life equivalent to four Titanic disasters would have been a more appropriate analogy.'

One of Britain's first responses to the impending threat posed by AIDS was to establish in February 1985 an advisory group of scientists, doctors and health officials. One of the group's first decisions was to tell the government that there was no need at that time to make the disease notifiable. The group's other roles included advising government about guidelines for medical staff dealing with AIDS patients, and promoting health education for the general public and those at risk of AIDS. It also advised on the screening of blood for the AIDS virus and ways of dealing with the problem of contamination of factor VIII, the clotting agent needed by haemophiliacs.

The advisory group quickly warned the government that it needed to spend more money on fighting AIDS. By September 1985, the Department of Health realized that AIDS was beginning to burn a large hole in the coffers of health authorities in London, which had the heaviest load of AIDS patients. The department allocated an extra £1m, mainly for London, and promised to investigate the possibility of launching a national campaign to educate the public about AIDS.

At the end of 1985 the Secretary of State for Social Services, Norman Fowler, announced that he was preparing a 'package' of measures to tackle AIDS, including an information campaign costing £2.5m. The money would pay for a campaign that was to begin in the spring of 1986 and run throughout that year. Fowler said: 'The campaign will be directed at the public in general and this will be coupled with a series of targeted campaigns for those known to be at special risk. The aim is to improve understanding of the disease and the ways in which its spread can be controlled.'

Scientists were still not convinced, however, that the government was really doing all that it could at this time. Professor Julian Peto, a cancer specialist at the Institute of Cancer Research in London, told the *Observer*: 'The country is just sitting back waiting for half a million people to be infected, instead of the 10,000 or so that we have at present . . . We are heading inexorably towards an AIDS crisis like the one in America today.' At that time, at the end of 1986, the government had just introduced screening of blood donations, and many organizations were concerned that the government was not spending enough money on adequate counselling of people who proved positive for infection with the virus. David Miller, a psychologist and counsellor for the hospital with more AIDS patients than any other in Britain, St Mary's in London, said: 'Like everyone else trying to deal with AIDS, we are having to compete fiercely for funds. It is a real uphill battle . . . The longer it takes to get that money, the further behind we get and the more unnecessary cases of AIDS develop as a result.' Jonathan Weber, another scientist from the same hospital, summed up the feelings of many professionals having to cope with the growing problem of AIDS at that time: 'Despite repeated warnings, the government has shown no intention of planning for the future.'

By the time the General Election had arrived, in the spring of 1987, the British government had become wise to this persistant criticism. It had already instigated a national advertising campaign, with commercials on television and the press warning people not to 'die of ignorance'. And just before the election, and a day before a by-election, the government announced that it was giving £14.5m to the Medical Research Council specifically for a 'directed programme' of scientific research into AIDS. It was a welcome sweetener for the council, which could not remember the last time that the British government had given medical research all that it had asked for. In less than three years, the government's attitude to funding research into AIDS had changed. Asked at the end of 1984 whether the government would allocate additional funds for research into AIDS through the Medical Research Council, Peter Brooke, a junior minister, replied: 'It is for the Medical Research Council to decide on its scientific priorities within the resources available to it.' At that time, the council had less than £500,000 to spend on AIDS research. Three more years into an epidemic, and an impending General Election, had at least begun to change this. Science in Britain had finally received the tonic it needed to enter the battle against AIDS. In France and America, scientists had already been working for quite some time on one of the most intriguing mysteries of AIDS. They were already searching for the virus.

CHAPTER 3

RACE FOR DISCOVERY

When scientists first discovered AIDS, they knew nothing of what was really behind the handful of diseases that eventually killed the sufferers. Initially, all they knew was that the people with AIDS fell into recognizable 'risk groups', such as homosexuals or drug users who injected their addiction. All these patients had one thing in common – their immune systems, the body's mechanism for fighting off infections, had all but collapsed. One particular white blood cell, called the T-helper cell, or 'T-cell', is an important part of the immune system. In AIDS patients, scientists found that this blood cell had practically disappeared from the body. Without the T-cell, the body becomes incapable of destroying the many different types of disease that constantly threaten good health. In 1981, scientists could only guess the reasons why the immune systems of people with AIDS should be suppressed, or weakened, in such a way.

One suggestion at the time was that AIDS was the result of the lifestyle that some gay men led in the liberated climates of some American cities, such as San Francisco and New York. Perhaps AIDS was caused by a bad batch of 'poppers', stimulants of amyl nitrites that some gay men took to give them a high. The first report of *Pneumocystis* pneumonia in five homosexual men, published in the *Morbidity and Mortality Weekly Report* in June 1981, noted that 'all five reported using inhalant drugs'. Scientists wondered, therefore, whether this could be the cause of the new epidemic. Researchers from the Centers for Disease Control collected batches of 'poppers' to analyse – but they found nothing that could indicate why such drugs should cause AIDS although poppers did appear to suppress the immune system.

Another observation pointed to something else. Very early on in their study of AIDS, researchers from the CDC discovered that gay men who had thirty or more partners a year seemed to be far more likely to develop AIDS than gay men who had one or only a few partners. Perhaps, therefore, these more promiscuous gay men had somehow overloaded their immune systems. The theory suggested that these men were more likely to have suffered from a wide variety of sexually transmitted diseases. This constant battering of the immune system, so the idea went, meant that the T-cells became seriously 'overworked' and depleted. Doctors had noticed soon after the first few cases of AIDS appeared that gay men who had multiple partners, but without the symptoms of AIDS, had suppressed immune systems. Many had a variety of sexually transmitted diseases. Perhaps, therefore, repeated 'inoculation' of semen during sex with different men had led eventually to the complete breakdown of the immune system, resulting finally in the infections typical of AIDS.

This theory could also apparently explain why some intravenous drug users, both men and women, were developing AIDS. These people were likely to use dirty needles and syringes, which meant that foreign particles entered the blood along with the drug. Often the drug itself was also impure. However plausible, the 'overload' theory could not explain why in the summer of 1982 the CDC discovered another group of people 'at risk' of developing AIDS. These were haemophiliacs, males with an inherited disorder of the body's mechanism for clotting blood. (Female haemophiliacs are very rare.) Most haemophiliacs have to inject regular quantities of factor VIII, a substance they lack in their blood. Factor VIII is necessary for blood clots to form. Haemophiliacs who do not get factor VIII when they need it can literally bleed to death. These males did not fit into the overload theory, however. Most were neither promiscuous homosexuals nor users of illicit drugs, and some were pre-pubescent boys.

More than 350 cases of AIDS were reported in the US by the summer of 1982, of which a handful were haemophiliacs, none of whom had the recognizable risk factors associated with the disease. The inclusion of haemophiliacs quickly supported the idea that the disease was caused by an infective agent, probably a virus. As soon as the CDC had published the first accounts of AIDS, medical scientists suspected that a virus could be the cause of the immune suppression. The trouble was trying to prove that any one of hundreds of viruses was the cause – and there was always the possibility that it could be a completely new virus.

The fact that haemophiliacs were now succumbing to AIDS strongly suggested that a virus was to blame. Factor VIII is made by pooling the blood serum of up to 30,000 donors and then extracting factor VIII from this pool. During the process of purification, the blood serum is filtered, which ensures that relatively large contaminants, such as bacteria and fungi, do not foul the end product. But smaller particles, such as viruses, survive. Perhaps, therefore, a virus was the cause of AIDS, and body fluids, such as semen and blood, passed it from one person to another. The CDC found that haemophiliacs with AIDS were distributed widely throughout the US, and not just concentrated in certain cities as were the homosexual sufferers of the disease. This also suggested that the infective agent was indeed lurking in batches of factor VIII sent to different parts of the US.

Further research on tracing partners of gay men with AIDS pointed to an infectious agent. The Centers for Disease Control found in 1982 that several gay men from different parts of the US had developed AIDS. They had all had sex with the same man, who also had AIDS, but not with each other. This was strong evidence that homosexuals suffering from AIDS could pass it on to other gay men. Doctors found yet more evidence that a blood-borne agent caused AIDS when they noticed that some men and women who had received blood transfusions had also developed AIDS. A twenty-month-old baby from San Francisco had developed AIDS after receiving blood transfusions immediately following birth. The doctors traced the nineteen donors who had contributed to the batches of blood and blood products that the baby received. One of these donors had developed AIDS ten months after donating blood.

At the beginning of 1983, the Centers for Disease Control found that women who were the sexual partners of men with AIDS also developed the disease. Some of these women had become pregnant and doctors found that they had passed AIDS on to their babies. By now it was clear that AIDS was caused by an infectious agent. The exact nature of this agent was still a mystery, however.

Virologists around the world began to take an increasing interest in this strange disease. If they could discover a virus responsible for AIDS they would receive much kudos, and possibly a Nobel prize. In 1982, several scientists suggested that AIDS could be linked with known viruses. One theory was that the disease resulted from a new strain of African swine fever virus, which infects pigs. This virus was decimating the pig population in Haiti at about this time, and Haitians appeared to be yet another group at special risk of AIDS. (The risk groups seemed to have just one thing in common, the letter 'H' – homosexuals, heroin addicts, hookers, haemophiliacs and now Haitians.) The theory was that the swine fever virus had mutated and infected humans. It fitted in nicely with the knowledge that many gay men from the US spent their holidays in Haiti. Perhaps they had picked up the virus there by having sex with locals who had themselves become infected by eating contaminated pork regularly, or by having contact with infected pigs. These Americans had then taken the virus back to New York and San Francisco.

The idea also suited a grander conspiracy theory. African swine fever virus first appeared in Cuba, Haiti's neighbour, in the mid 1970s. Cuba claimed at the time that the CIA had deliberately infected the island with the virus in order to undermine the local economy, which is heavily dependent on pig farming. From Cuba, this virus then spread to Haiti. Perhaps the CIA had tinkered with the virus in a way that allowed it to cross the species barrier and so infect humans as well as pigs. Could the CIA be responsible for AIDS? It was not the first or indeed the last time that the CIA, or the KGB, was linked with the disease. For most scientists, however, the idea had little currency, just as there was no evidence for the CIA or KGB making the virus in a laboratory during a germ-warfare experiment that went wrong.

In 1982, the race to find the cause of AIDS had sparked the interest of a relatively small group of scientists working in a specialized area of virology. One of these scientists was Robert Gallo, then the head of a laboratory at the US's National Cancer Institute at Bethesda, Maryland. Early in 1982 Gallo and other scientists suggested that AIDS might be caused by a type of virus called a retrovirus. Usually, viruses are made of a genetic material called DNA (deoxyribonucleic acid), just like that in humans. The viral DNA is wrapped in a coat of proteins. The virus attaches to the outer membrane of the cell it is attacking, and then injects its DNA into the cell. The cell treats the viral DNA as its own genetic material. The cell therefore uses the viral DNA as a blueprint to make viral protein. Eventually the cell manufactures vast quantities of viruses, which finally burst from the cell to infect other cells. Instead of DNA, retroviruses have another type of genetic material, called RNA (ribonucleic acid), which they inject into cells. On its own, this type of viral RNA is next to useless, so retroviruses also have an enzyme that can make DNA from RNA. In most living things, the DNA is copied into RNA, which then goes to make proteins. In retroviruses, the sequence is RNA to DNA, then back to RNA and then to proteins. This initial reversal of the usual sequence, from RNA to DNA, is why scientists describe these viruses with the prefix 'retro'.

In 1978, Gallo and a group of scientists from Japan had discovered for the first time a retrovirus which infects humans. It causes a type of leukaemia, called adult T-cell leukaemia, so Gallo called this virus human T-cell leukaemia/lymphoma virus (HTLV). (A lymphoma is a type of tumour.) In 1982, researchers in Gallo's laboratory found another retrovirus in humans, a much rarer virus that also causes a type of leukaemia. Gallo called this virus HTLV-2, to distinguish it from the first human retrovirus, HTLV-1. In the same year, Gallo suggested that a form of the HTLVs might well be responsible for the new and devastating disease of AIDS. His reasons were these: the HTLVs attacked T-cells, they were transmitted by blood and intimate contact, and Gallo had noticed that the immune systems of some leukaemia patients infected with HTLV-1 were slightly suppressed. Research into another retrovirus, which causes a form of leukaemia in cats, backed up Gallo's theory. These cats also sometimes had an immune deficiency.

Gallo was not alone. In Paris, researchers at the Pasteur Institute (a grand research institution established in 1887 by the great microbiologist Louis Pasteur) also began to investigate the idea. One hospital in Paris, La Pitié Salpétrière, had a patient suffering from the early symptoms of AIDS. This patient, a gay man, had a condition that doctors describe as lymphadenopathy, or swollen lymph glands. These glands, in the groin, armpits and neck normally contain plentiful T-cells. The doctors treating this patient took a small piece of tissue from his lymph glands and, on the morning of 3 January 1983, sent it for analysis to virologists at the Pasteur Institute.

The researchers, led by Luc Montagnier, were particularly interested in patients with the early symptoms of AIDS because patients with full-blown AIDS had too few T-cells for the investigators to analyse properly. This patient offered enough T-cells to grow and to study. Montagnier's assistants, Françoise Barré-Sinoussi and Jean-Claude Chermann, began to grow the cells in the laboratory to see whether they could spot signs of a retrovirus. Barré-Sinoussi looked for the presence of a chemical unique to these viruses, called reverse transcriptase. The procedure was a standard one. The researchers first separated the T-cells from the rest of the tissue by spinning the tissue in a centrifuge. (They then added anti-interferon, a substance capable of neutralizing the natural interferon in the tissue sample, because interferon seems to inhibit the growth of retroviruses.) Following this they added a chemical that stimulates the growth of T-cells, called T-cell growth factor, along with other chemical stimulants. (Gallo was one of the scientists who had discovered T-cell growth factor in the mid 1970s. He had given the chemical to researchers at the Pasteur.)

Every three or four days, the researchers looked for viral activity. They did this by taking a sample of the culture, and spinning it at high speed to concentrate any viruses as a pellet at the bottom of the test tube. They treated the pellet with a detergent, which would open the protein coat of retroviruses to release the reverse transcriptase inside – the chemical that signals the presence of a retrovirus. On 25 January 1983, the researchers at the Pasteur Institute for the first time found evidence of reverse transcriptase. On this day they had added a number of chemicals, such as RNA and the simpler chemicals that can be made into DNA, and found that molecules of DNA formed. Reverse transcriptase was indeed present; it was converting simple molecules into longer chains of DNA. This indicated for the first time that the cells from the patient were harbouring a retrovirus. Barré-Sinoussi carefully recorded the event in her laboratory notebook. The activity, she wrote with the caution of an experienced scientist, was very low and so the evidence was inconclusive at this stage.

Two days later, she and Chermann looked again. The activity, measured by the amount of DNA that was made, had increased, reaching a peak on 7 February. The results looked exciting, but, on 11 February, the researchers detected that the activity had begun to fall 'which both surprised us and made us anxious', Barré-Sinoussi said later. The only other known retrovirus to infect humans at that time, Gallo's HTLV-1, had exactly the opposite effect. It caused the T-cells to multiply uncontrollably, just like a cancer, so that there is an ever-increasing amount of reverse transcriptase. This new virus seemed to be killing off the T-cells that it infected. Barré-Sinoussi later on said that 'this was our first suspicion that the virus we had was not like HTLV'. Meanwhile, Gallo and others continued to suggest that the HTLVs were the cause of AIDS.

The group at the Pasteur Institute had come across one of the peculiarities of this virus that made it so difficult to study. It grew in T-cells, but instead of causing the T-cells to multiply, so providing more fodder for the virus to infect, and thereby increasing the amount of virus to study, the virus destroyed its own habitat. Researchers had a tough time trying to produce enough virus to analyse in any detail. The French group, in those first few months of 1983, had a solution. They kept adding fresh T-cells from a healthy donor to provide a constant source of new cells within which the virus could replicate. After trying this for two weeks, Montagnier, Chermann and Barré-Sinoussi found they could increase the amount of virus they had. Furthermore, they could take small quantities of the fluid surrounding the T-cells and infect other T-cells with the virus. This showed that the virus existed outside the cells.

In virology, as in most spheres of life, seeing is believing. The French group quickly tried to photograph the virus under very high magnification. On 4 February 1983, Charles Dauguet, an electron microscopist at the Pasteur Institute, was the first person to see pictures of the virus. The images were of poor quality because the subject matter was so difficult to photograph. At the end of March and the beginning of April he took better photographs. The group from the Pasteur Institute was satisfied that the pictures were of a virus, and decided to publish the results. Before doing so, however, they wanted to see whether the virus they had found bore any resemblance to the only other known retrovirus to infect humans – HTLV (HTLV-2 was still not well known). If the French virus was not the same, then it must be a new virus. Montagnier's group asked Gallo to send samples of T-cells infected with HTLV, and also certain chemicals, called antibodies, that will bind specifically to HTLV. The body produces antibodies to fight off foreign particles. Scientists make antibodies by injecting viral proteins (in this case the proteins of HTLV) into animals. The immune systems of these animals then make antibodies that attack these foreign proteins. If the virus that Montagnier's group had just found was HTLV, or even similar to this virus, then the researchers would have found that the HTLV antibodies and the new virus's own proteins would bind together like a lock and key. Virologists call this 'cross-reactivity'.

The team from the Pasteur Institute found no cross-reactions. The new virus seemed to be quite unlike HTLV. The only structural similarity appeared to be that one of the new virus's proteins was a similar size to a protein found in HTLV. How did the new virus fit in with the existing family of retroviruses? Clearly there was some link with HTLV. Both viruses were retroviruses that infected humans for a start, but how closely related were they?

Montagnier wrote a scientific paper describing these results, and sent it to the American journal *Science* for publication. *Science* asked Gallo to read the paper and make comments – it is common practice in scientific research for competing scientists to referee each other's papers before publication. Gallo was unsure whether Montagnier had really found a new retrovirus. If he had, then Gallo thought that the virus must be similar, if not the same as HTLV. 'Publish,' he told *Science*, but he advised Montagnier to make alterations to his paper and make it clear that his 'new' virus was closely related to HTLV. Montagnier took Gallo's advice and wrote: 'We report here the isolation of a novel retrovirus from a lymph node of a homosexual patient with multiple lympha-denopathies. The virus appears to be a member of the human T-cell leukaemia virus (HTLV) family.' The phrase would come back to haunt Montagnier. He was about to become embroiled in an ugly public row with Gallo over, among other things, the correct name for the virus. His researchers firmly believed that they had found something quite unlike HTLV. The wording of this paper gives an indication of how the scientists at the Pasteur Institute themselves seemed to be pulled in different directions. This first scientific paper to describe the AIDS virus, published on 20 May 1983, opens with the statement: 'A retrovirus *belonging* to the family of

32

recently discovered human T-cell leukaemia viruses (HTLV), *but clearly distinct* from each previous isolate, has been isolated from a Caucasian patient with signs and symptoms that often precede the acquired immune deficiency syndrome (AIDS)' (authors' emphasis). It might seem odd that something can belong to the same family of something and yet be clearly distinct from all of the known members of that family. The similarity between Montagnier's new virus and Gallo's HTLV was that they were both retroviruses, and they both infected the T-cells of humans. Nevertheless, the team from Paris made it clear that they had discovered a 'novel' virus.

The French researchers did not name the new virus. They just described it as 'lymphotropic', meaning that it had an affinity for the white blood cells known as lymphocytes, the T-cells in this case. Montagnier and his team were a long way from proving that their new virus actually caused AIDS. They had, after all, only isolated it from one patient with the early signs of AIDS: 'The role of this virus in the etiology of AIDS,' they wrote, 'remains to be determined.' Nevertheless, the paper did show, for the first time, relatively clear photographs of the virus. It was undoubtably a breakthrough in the race to find the cause of AIDS.

Unfortunately for Montagnier, the scientific world virtually ignored the discovery described in his paper. One of the reasons was that three other scientific papers on AIDS dominated the same issue of *Science*. All described a link between AIDS and the HTLV virus discovered by Gallo. Two papers came out of Gallo's laboratory, and the third came from the laboratory of a close collaborator and friend of Gallo, Myron (Max) Essex of the Department of Cancer Biology at the Harvard School of Public Health in Boston. Between them, Gallo and Essex put a convincing case for the view that HTLV caused AIDS. Gallo published photographs of HTLV viruses found in an AIDS patient. He said that the patient had antibodies to HTLV proteins. He added that the relative absence of AIDS in Japan, where HTLV-1 has been prevalent for many years, may be because the people there had developed a resistance to the virus.

Montagnier's results did not fit nicely with the papers of Gallo and Essex, which became accepted wisdom. Science itself was unsure of the significance of Montagnier's findings. In the same issue, a reporter for the journal described Gallo's research in detail. Montagnier's work merited one sentence. Science, the journal that published the first description of the AIDS virus, failed to notice the true importance of the discovery. The world would have to wait a little longer to realize the importance of Montagnier's virus.

In 1987, Robert Gallo wrote of Montagnier's 'intriguing' discovery: 'The initial report . . . was hardly a conclusive identification of the cause of AIDS.' He was right. It was no good finding a virus in one patient; the team from the Pasteur Institute had to find it in many more people with AIDS, or with the early symptoms of AIDS. In the summer of 1983, Montagnier's group set about this task. The researchers concentrated on developing a test to see whether people suffering from AIDS were infected with their new virus. The aim was to compare the test results of these people with those of a control group of people not at risk of developing AIDS. If the people with AIDS were infected with the virus, but the control group was not, then the researchers were on their way to establishing a causal link between the virus and the disease.

The principle of the test was relatively straightforward. People with the virus should have antibodies to it. Therefore, if blood serum is mixed with the virus in question, or with cells containing the virus, any antibodies in the blood should identify and stick to the virus. With a few clever tricks this shows up visually – as a colour change, for example (see pp. 72–4). The problem that the group in Paris faced in the summer of 1983 was that the researchers had found it difficult to grow enough virus to make such a test. The obstacle was a familiar one: this virus quickly destroyed the cells in which it lived. The researchers had to rely on continually replenishing the T-cells in which the virus grew, a messy and unsatisfactory process.

As the French researchers came closer to a test, the administrators at the Pasteur Institute became aware of the financial benefits of developing such a blood test. The test would have to be properly patented to protect it against copying. The institute therefore deposited a sample of the new virus at France's National Collection of Cultures of Microorganisms on 15 July 1983. This is a necessary step in the patenting process. The institute, on Montagnier's advice, called the virus lymphadenopathy AIDS-associated virus (LAV), to distinguish it from other viruses, especially the HTLVs. Two days after making this deposition, Montagnier sent a frozen sample of the virus to Gallo's laboratory. The sample was damaged on arrival, so the team in the US failed to grow any virus from it. Gallo and his researchers were in any case preoccupied with the data they were working on showing a link between AIDS and the HTLV viruses.

Work in the laboratories of the Pasteur, meanwhile, went on at fever pitch. Barré-Sinoussi and Chermann wanted to develop a reliable test quickly and to publish their results as soon as they could. During those summer months they became convinced that their LAV was the cause of AIDS. Now they wanted to convince the sceptical world of virology as well. In September, they had their chance. More than a hundred virologists from all over the world met at Cold Spring Harbor Laboratory in New York to discuss human retroviruses. It was a closed scientific conference where reporting was not allowed. Gallo was one of the organizers, and he invited Montagnier to present his latest research.

One of the virologists at this meeting, Don Francis of the Centers for Disease Control, who listened to Montagnier's presentation, said afterwards that it was clear that the Frenchman had a totally new human retrovirus, quite unlike the HTLVs. His pictures were of a virus with a cone-shaped core. The core in the HTLVs was cylindrical. More importantly, Montagnier could now link his new virus directly to AIDS. His preliminary results with his test, he told the conference, showed that he could identify antibodies to the LAV virus in 22 of 35 patients suffering from swollen lymph glands, or lymphadenopathy. He could also detect antibodies to LAV in 7 out of 40 healthy homosexual men, and antibodies in one out of 54 control samples of blood from healthy heterosexuals. Montagnier said he had 'conclusive' evidence that LAV represented a separate group of human retroviruses, quite distinct from the HTLVs.

At this stage, Montagnier's test was clearly not 100 per cent accurate. The test was most probably wrongly identifying antibodies to LAV when there were no antibodies, and failing to identify them when they were present: so-called 'false positive' and 'false negative' results. This is probably why, for instance, his test could not identify the presence of LAV in all thirty-five patients suffering from swollen lymph glands. Nevertheless, the data looked impressive, though not to Gallo. The scientific journal *Nature* later described the atmosphere following the encounter between Gallo and Montagnier: 'It seems common ground between Gallo and Montagnier that the former was unreasonably dismissive of the latter at the end of that talk; far from being applauded for his perceptiveness, Montagnier was left with the impression that he had been scorned.' Gallo still evidently believed that HTLV or a variant of it caused AIDS. The name of the book of papers from the conference at Cold Spring Harbor reflected this: *Human T-Cell Leukaemia/Lymphoma Virus, the family of human T-lymphotropic retroviruses: Their role in malignancies and association with AIDS.* The book still labelled LAV as belonging to the family of HTLVs. Gallo and Max Essex were co-editors.

During the meeting at Cold Spring Harbor, Montagnier – despite his hurt feelings – agreed to give Gallo a second sample of LAV. He did so on 23 September 1983, a week after the Pasteur Institute filed for a patent in Britain. Montagnier made one of Gallo's researchers, Mika Popovic, sign a statement promising that Gallo and his group would not use the virus for commercial gain. The sample of virus was for research purposes only. In the event, according to Gallo, the virus from Paris refused to grow in sufficient quantities to be of any good, and so was useless. As events unfolded, the

Pasteur Institute would later begin to doubt this version of events.

At the end of September 1983, Robert Gallo seemed to become less convinced that he was really on the right track with HTLVs. He wrote to a colleague in West Germany, Professor Friedrich Deinhardt, a leading virologist from the Max von Pettenkofer Institute in Munich, to voice his opinions, and, more importantly, for Deinhardt to pass on these thoughts to other virologists in Europe:

'Dear Fritz,' Gallo wrote,

> After a recent trip to Europe I have become concerned that some people are under the impression that I believe AIDS is caused by HTLV. I am writing to you because of your central position in viral oncology [the study of cancer] in Europe, and I hope you will help me to dispel this impression when it comes up. My opinion is simply this: of the known candidates, it's a pretty good one based on conceptual grounds. When it comes to data proving the point, the results are stimulating for more studies, but clearly not definitive for any one virus . . . In my opinion an HTLV variant is the most likely candidate, and if it isn't this, it is an as yet unknown virus.

Gallo went on to describe research showing that HTLV appears to cause immune suppression. He then commented on the work of Luc Montagnier, whose latest research he had listened to just a fortnight previously, and whose virus was now sitting in a freezer in his laboratory: 'I have never seen the virus that Luc Montagnier has described, and I suspect he might have a mixture of two. On the other hand, some of his data are interesting but still far from definitive.' Even though Gallo was not now committing himself fully to the idea that HTLV caused AIDS, he was equally convinced that Montagnier was barking up the wrong tree.

The researchers in Gallo's laboratory faced the same problem as the team in Paris: all attempts to grow the virus

37

in sufficient quantities for analysis were thwarted because the virus killed the cells in which it lived. Then, in early November, Mika Popovic in Gallo's laboratory made a remarkable breakthrough. He tried to grow the virus in a certain strain of T-cells that scientists had discovered in 1979. Rather than killing the T-cells, the strain multiplied continuously. This strain was called HUT-78, and the particular subgroup that Popovic used was called H9. The effect of infecting this strain of lymphocyte was exciting. Suddenly, Gallo and his researchers could begin to harvest huge amounts of the virus that they suspected might cause AIDS. They had little problem then with developing a rudimentary test to see whether they could directly link the virus to AIDS.

Gallo collected blood samples from many different people suffering from AIDS in order to see whether he could detect antibodies to his virus. In the first few months of 1984, other researchers also began to get closer to the cause of AIDS. In Britain, Abraham Karpas, of the University of Cambridge, had found a virus in an AIDS patient attending a local hospital. He published a short paper describing it, along with photographs of the virus taken at very high magnification. At the University of California at San Francisco, a team led by Jay Levy was also studying AIDS and had begun to find evidence of a new virus. And at the Centers for Disease Control, Paul Feorino and co-workers were also on the trail of isolating the AIDS virus. Gallo's team had to work fast to avoid being pipped at the post by other scientists. They knew that finding a virus was one thing, but proving that it caused the disease was another matter entirely.

At another closed scientific conference, this time held at Park City in Utah in early 1984, Jean-Claude Chermann from Montagnier's team gave more details of LAV. Gallo was also present. Chermann believed that his data must surely vindicate the belief that LAV was the cause of AIDS. After his presentation, some scientists agreed with Chermann that they had now seen conclusive evidence that LAV caused AIDS. Within a few weeks of the meeting at Park City, Gallo called Don Francis, then the head of the virology section of the

AIDS programme at the Centers for Disease Control, to tell him that he had also found a virus. Francis immediately wanted to compare Gallo's new virus with LAV. To do this, Francis wanted to test several hundred blood samples taken from people either developing AIDS or with full-blown AIDS to see whether they contained antibodies to either Montagnier's virus or to Gallo's virus. These results would then be compared against blood from a group of people without AIDS, and not at risk of the disease – the control group. It would be the most definitive test to discover whether science could link either, or indeed both, of these viruses to AIDS.

Francis wanted Gallo and Montagnier to use their respective viruses to test the first batch of blood samples from the CDC, and for the CDC to use both viruses for its own confirmation. At this point, cooperation broke down, however. Francis said later that he detected a growing feeling of antagonism from Gallo. The results of this study were never published in full. Gallo was instead concentrating on publishing his own results. In April news of a 'variant' of HTLV began to leak to the American press. An article appeared in the *Wall Street Journal*, followed by the *Washington Post* and the *San Francisco Chronicle*. Soon after these reports, *New Scientist* in Britain reported that Gallo had found a 'third variant' of the HTLV family. The story came from a freelance journalist who had interviewed Gallo.

On 23 April 1984, before Gallo had the chance to publish his research in the normal way – in the scientific journals – the US Department of Health and Human Services, the ultimate paymaster for Gallo's laboratory, decided to hold a press conference in Washington DC to announce the 'discovery' of the AIDS virus. The same morning, lawyers from the US government filed a patent on a test for antibodies to the virus, developed by Gallo. Margaret Heckler, then the Secretary of the US Department of Health and Human Services, took charge of the press conference, despite her sore throat. 'First,' she said, 'the probable cause of AIDS has been found – a variant of a known human cancer virus, called HTLV-3.' (It is true that HTLV-1 and 2 are both cancer viruses, but it is now known that the AIDS virus is not a cancer virus.) 'In particular, credit must go to our eminent Dr Robert Gallo,' Heckler continued, 'who directed the research that produced this discovery.' A statement from the US government cultivated the chauvinistic tone of the press conference: 'Today, we add another miracle to the long honour roll of American medicine and science.'

Gallo thanked the other members of his group for their part in the discovery of the new virus. The only time that Montagnier's work was mentioned was when reporters asked about how Gallo's HTLV-3 compared with LAV. Gallo replied: 'If it [the virus] turns out to be the same I certainly will say so and I will say so in a collaboration [sic].' He added: 'I think the two laboratories are very likely to come together although I cannot say at this point whether the two viruses are identical.'

A few weeks later, on 4 May 1984, the full details of Gallo's research emerged in Science. Gallo described how he could produce the virus in massive quantities in the H9 cells. He could identify the presence of the virus in 48 out of 167 people at risk of AIDS, but found no evidence of the virus in 115 healthy heterosexuals. It was the most convincing evidence yet published that a single virus led to AIDS. Gallo, as Heckler said at the press conference earlier, named his new virus human T-cell leukaemia virus type-3 – HTLV-3. He evidently believed that the virus belonged to his own family of human retroviruses. In the eyes of the American media, Gallo was the discoverer of the AIDS virus. In Paris, however, the Pasteur Institute would plan to try to convince the world that the claim to the discovery belonged to Montagnier.

CHAPTER 4

BATTLE FOR CREDIT

On 15 May 1984, soon after Robert Gallo published the research showing that his HTLV-3 caused AIDS, two scientists, Fred Murphy and Jim Curran from the Centers for Disease Control, went to visit him at his laboratory in Bethesda, Maryland. The CDC wanted, among other things, to establish the relationship between the French virus, LAV, and HTLV-3. The CDC had a problem on its hands. There now seemed to be two causes of AIDS, the virus that Gallo had just announced, and the virus that Montagnier had discovered a year earlier. Murphy and Curran wanted an agreement with Gallo that would allow the CDC to experiment with samples of HTLV-3. They already had samples of LAV from Luc Montagnier's laboratory in Paris.

In a memo to his boss written a month later, Murphy described the meeting, which took place in Gallo's office. Gallo had a document that he wanted the two CDC scientists to sign before they took samples of HTLV-3 away with them.

As Jim and I have stated, it was a tense moment, fraught with the possibility of non-delivery. Our tack, stated orally in several different ways as we discussed the matter with Dr Gallo, was that public-health purposes were paramount. Dr Gallo agreed. In our conversation, it became clear that comparison of his HTLV-3 prototype with the French prototype LAV occupied a separate niche – the comparison was seen as having both academic and public-health purposes. Because of the latter, I offered, using several tacks, to have certain comparative tests between his HTLV-3 and the French LAV done at CDC; Dr Gallo declined each time, stating that such work would be done in his lab. It was clear from our discussion that this was the only subject which engendered such difficulty – when we switched to other themes . . . there was no problem.

The written agreement that Gallo wanted seemed to be a standard one, except for a 'seventh item', according to Murphy, which appeared to be 'for CDC only'. Gallo had given collaborators in other laboratories samples of HTLV-3, but since the CDC could be considered to be a competitor, Gallo told Murphy, a restriction would have to be placed on the use made of his virus. 'The [seventh] item stated that CDC was prohibited from using the material from [Gallo] for comparison with other viruses [taken to mean LAV or surrogate for it],' Murphy wrote in his memo. Gallo told Murphy and Curran that he was putting this restriction on the use of HTLV-3 so that his own researchers could capitalize on their basic discoveries.

Everyone now wanted to know the relationship between LAV and HTLV-3. The scientific community was confused with the notion that two viruses, with two different names, could cause the same disease. Did they both cause AIDS? Were they the same virus? In the CDC's preliminary studies of about two hundred blood samples, using primitive blood tests for antibodies to LAV and to HTLV-3, the CDC had little doubt that they were the same virus. But detailed proof, by analysing the molecular structure of the two viruses, had yet to emerge. Such analysis took time to do well.

It is possible to compare viruses in a number of ways – looking to see if antibodies match up is one way, but this only gives a crude measure of how related two viruses are. A more accurate method is to search for the presence of particular sites along the virus's genetic material, or DNA. Scientists do this by adding a number of different chemicals to the DNA.

These chemicals, called restriction enzymes, cut the DNA only when the enzyme recognizes a specific site. Each type of restriction enzyme identifies a different site. The process is like snipping a necklace of beads. Imagine that each bead bears a letter from the alphabet, and that the beads are arranged in a random sequence: some restriction enzymes cut the necklace only between beads bearing the letters E and P, say, and others only between other pairs of letters. After subjecting the 'alphabet necklace' to a range of restriction enzymes, you will end up with a number of smaller segments of the necklace of differing lengths. If the sequences of letters on the beads of two necklaces are very similar or identical to begin with, then two batches of the batch of smaller necklaces you end up with will also be similar. If the sequences of two necklaces are different, then so are the final batches of smaller necklaces (see figure 1).

The DNA of viruses (and remember that retroviruses, such as the AIDS virus, make a DNA copy of their RNA, and scientists prefer to analyse retroviruses by studying the viral DNA rather than viral RNA) is like a necklace of beads. Mix the chain of viral DNA with a number of different restriction enzymes, and you end up with a group of smaller strands of DNA, of differing lengths. If the sequences of two chains of viral DNA are very similar, or identical – showing that the two viruses are the same – then you end up with two very similar or identical groups of smaller chains of DNA. Molecular biologists call this 'restriction mapping', and it is a relatively quick way of seeing how related two viruses are (see figure 1).

The problem with restriction mapping is that it does not tell you anything about what is between the particular sites along the DNA that the restriction enzyme identifies. If you have an enzyme that identifies only E and P, what about the other twenty-four letters that may be in the 'alphabet necklace' between Es and Ps? It is possible for scientists to work out the detailed sequence of DNA, in the same way that it would be possible to take one bead after another from our necklace and make a note of its letter. This takes much longer

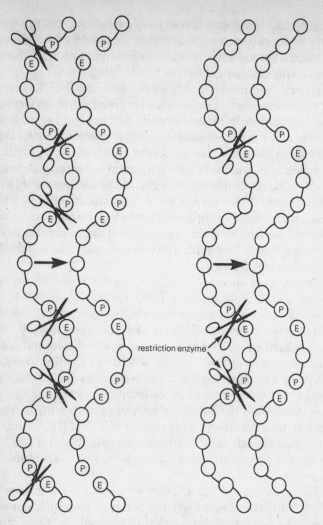

Figure 1. Special chemicals called restriction enzymes cut the genetic material of the virus at certain sites, labelled P and E here. The sites may or may not be present, depending on the strain of the virus. Different strains therefore produce different lengths of genetic material after the treatment, which is called restriction mapping. In other words, different strains of virus have different restriction maps.

than restriction mapping, but 'molecular sequencing', as the process is called, is a definitive method of identifying the precise relationship between two bits of DNA, and therefore of working out how closely related two viruses are.

During 1984 Robert Gallo's team in the US and Luc Montagnier's colleagues in France were busy mapping and sequencing their respective viruses. They published the full sequences of their viruses in January 1985 within days of each other: the French published in the American scientific journal *Cell*, and the Americans published in the British journal *Nature*. The wider scientific community now knew what the smaller world of virology had suspected for some time: the two viruses were indeed one and the same. The genetic sequences were very similar – roughly 99 per cent identical. As one scientist at the Pasteur Institute remarked: 'The two viruses are not identical but they are indistinguishable.'

During that period, researchers in other laboratories also began to isolate and to grow samples of viruses that they had found in people with the early symptoms of AIDS, or with the full-blown disease. Jay Levy of the University of California at San Francisco had found a virus that he called AIDS-related virus, or ARV. This too seemed to be strongly connected with AIDS. How did ARV relate to HTLV-3 and to LAV? Two researchers from the National Institute of Allergy and Infectious Diseases, which is next door to Gallo's laboratory in Bethesda, wanted to find out. Arnold Rabson and Malcolm Martin came to an astonishing conclusion: although LAV and HTLV-3 were practically identical, the molecular sequence of ARV was substantially different. Later on, other researchers also began to compare AIDS viruses found in different people from different parts of the world. The aim was a serious one: scientists wanted to find out how variable the virus was. This would have important repercussions in the fight to develop a vaccine. The more the virus varies, the harder it would be to make a vaccine.

Researchers from the Pasteur Institute also compared their LAV with HTLV-3, as well as with the ARV virus from California, and with two viruses found in people living in Zaire in Africa. The chemical sequences of LAV and HTLV-3 differed by about 1 or 2 per cent. This compared to differences of about 10 or 20 per cent or more with the other three viruses. The similarity of LAV and HTLV-3 appeared to be a fluke. The AIDS virus was much more variable than the two molecular sequences of LAV and HTLV-3, published in January 1985, first suggested. Researchers soon realized that it was not going to be easy to develop a vaccine that would protect people against all the different variations of this virus.

Scientists now know that the virus that causes AIDS is quite peculiar in the extent to which it can vary – in other words the rate at which it can mutate. In 1987, researchers at the Los Alamos National Laboratory in the US analysed the mutation rates of seventeen different AIDS viruses. They came to the conclusion that parts of the virus, for example the outer coating of the virus, called the envelope proteins, may be changing at a rate five times faster than that of the influenza virus. These scientists predicted that the structure of each AIDS virus is changing overall by as much as 1 per cent a year. This means that there may be a 2 per cent difference between two viruses a year after they came from a common ancestor. Parts of the virus, such as the outer envelope protein, can change by more than 30 per cent in a short period of time. One of the scientists from Los Alamos, Gerald Myers, said that the variation of the virus that researchers first began to detect in 1985 was not the complete story: 'It became clear that what we saw in 1985 in terms of variation was only the tip of the iceberg.'

One suggestion to account for this variation relates to the enzyme reverse transcriptase, which copies the virus's genetic material, its RNA, into the other type of genetic material, DNA. This enzyme is said to be 'unfaithful'. It does not make a perfect copy of DNA from RNA all the time. Every now and again, the enzyme lets a little error slip in. In other words, it is permitting the virus to mutate, and mutations cause changes in the structure of the virus. In terms of the analogy of the alphabet necklace, it means that reverse transcriptase is slipping in one letter when it should be inserting another into the sequence of chemicals. For a retrovirus, variety is not just the spice of life, it is the key to survival, because it means that the virus can adapt quickly to a changing environment. For a virus such as the AIDS virus which attacks the body's immune system this is a great advantage. Its rapid rate of mutation means that the virus can continually fool the body's immune system.

The discovery of the variation of the AIDS virus began to fuel deep suspicions within the Pasteur Institute about the extraordinary similarity of Gallo's HTLV-3 and Montagnier's LAV. Could Gallo have used Montagnier's sample of the virus, which Montagnier gave to Gallo twice in 1984, in order to help to make his own discovery of HTLV-3? If this was the case, then the blood test for antibodies to the virus that Gallo's laboratory had developed using HTLV-3 would represent a commercial use of the French virus. The written agreement between the Pasteur Institute and Gallo's laboratory had expressly forbidden this.

The Pasteur Institute applied for a patent in the US on Montagnier's blood test in December 1983. The US government applied for a patent on Gallo's blood test in April the following year. However, in May 1985, the US Patent and Trademark Office awarded a patent on Gallo's test, without making an award for Montagnier's earlier patent application.

The frustration felt by the scientists at the Pasteur Institute over what they believed was a failure to recognize their pioneering work eventually turned into an ugly public row involving armies of high-powered lawyers. In December 1985, the Pasteur Institute went to court. The institute wanted recognition for its scientists, and a share of the royalties that the US government was beginning to receive on the blood test developed by Gallo. Case number 730-85-C in the US Claims Court – the Pasteur Institute versus the US government – concerned one very important allegation. 'Upon information and belief', the Pasteur's lawyers claimed, the virus at the heart of the blood test developed by Gallo 'is, or is substantially identical to, the LAV strain first isolated by Pasteur'.

Before the dispute had got this far, Gallo had tried to explain how his HTLV-3 could be so similar to Montagnier's LAV. It might be, he said in a letter to the journal *Nature*, 'because the individuals from whom these isolates were derived acquired the virus at a similar time and place. In fact, many of our earliest HTLV-3 isolates were all from specimens obtained in late 1982 or early 1983 from the east coast of the United States, and LAV, although isolated from a Frenchman with a lymphadenopathy syndrome, had his contact in New York in the same period.'

In fact the last time the Frenchman in question had visited New York was in 1979 – two to three years before the blood samples in question were taken from gay men, and then sent to Gallo's laboratory. Gallo implied that the similarity of HTLV-3 with LAV was because his virus had come from a close sexual partner of the Frenchman who was the source of the Pasteur's LAV. This was why the two viruses were virtually identical.

This explanation had one problem, which only became apparent later on: as already mentioned, the virus mutates very quickly. Even if Gallo's blood samples had come from a close sexual partner of the Frenchman, the chances are that the virus would have changed quite considerably in two to three years. At the end of 1985, nearly a year after Gallo had offered this explanation for the similarity of the viruses, a team of scientists, led by Steven Benn and Rosamond Rutledge of the US National Institute of Allergy and Infectious Diseases in Maryland, published an analysis of twelve different AIDS viruses, including LAV and HTLV-3, along with five different isolates of the virus taken from people living in New York. They subjected the viruses to seven different restriction enzymes, which were known to cleave the viral DNA at different points. They found: 'With the exception of LAV and HTLV-3, all of the isolates were different.' They also discovered that the five viruses from New York were all different from each other, and different to LAV and HTLV-

3. It seemed, therefore, that the AIDS virus varied even within the same city. This did not fit well with Gallo's theory explaining why HTLV-3 was so similar to LAV.

The lawyers working for the Pasteur Institute, Townley and Updike, were eager to pounce on all this circumstantial evidence to show that Gallo's virus 'is, or is substantially identical to, the LAV strain' of Montagnier's group. The firm, which operates from offices occupying several floors of the Chrysler Building in New York City, had put a bright young lawyer, Jim Swire, on to the case. His brief was to dissect the difficult subject of virology and present the Pasteur Institute's case in plain and simple English. He was also Townley and Updike's tough guy, appointed to cope with the experienced and shrewd lawyers working for the US government. Swire once described himself as 'the hired gun' of Townley and Updike.

One of Swire's first tasks was to apply for documents from Gallo's laboratory under the US Freedom of Information laws. He wanted to know anything and everything that went on in Gallo's laboratory before and after Gallo announced the discovery of HTLV-3. He also obtained documents from other laboratories that had contacts with Gallo's lab. One of these was the Electron Microscopy Laboratory of the Frederick Cancer Research Facility in Maryland. This laboratory took high-magnification photographs, called electron micrographs, for Gallo's researchers, who did not have an electron microscope of their own.

Swire came across a letter from the head of the Electron Microscopy Laboratory, Matthew Gonda, to Mika Popovic, the researcher in Gallo's laboratory who first successfully grew HTLV-3. The letter, dated 14 December 1983, contained the results of analysing thirty-three samples of blood that Popovic had sent to Gonda. The letter said that just two of these samples proved to be positive for 'lentivirus' (lentiviruses are a type of retrovirus, and the AIDS virus belongs to this group).

This was a clear sign that Popovic had found the AIDS virus, and was growing it in culture. What interested Swire, however, was that Gonda had referred to each of these samples as 'LAV', because this is how these two samples were labelled when they reached his laboratory. Swire suspected that Gallo's laboratory was growing the French virus, LAV, even though Gallo had vehemently stated that the sample of LAV he received from the Pasteur Institute failed to be of any use. Gallo says that at this time there was no other name for the AIDS virus but LAV. It was hardly surprising, he said, that his researchers labelled the virus 'LAV'.

Swire's suspicions were aroused further when he received an anonymous tip-off about an illustration that Gallo published in one of his papers describing the discovery of HTLV-3. This illustration comprised three rows of three photographs – nine pictures in all. Each row of three pictures depicted one of the three HTLV viruses in different stages of development: HTLV-1, HTLV-2, and the AIDS virus, HTLV-3. And this same photograph was distributed to reporters at the press conference to announce HTLV-3. Gallo wanted to show the similarity of HTLV-3 to the other two viruses. Unfortunately, the photograph was in fact of the French virus, LAV. There had been a dreadful mistake, and two years after Gallo first published these photographs, he had to correct his error by writing a letter to the American journal *Science*, which had first published the pictures.

Gallo later said in an interview in *New York Native*, a gay newspaper, that the mistake was a 'sloppy, embarrassing mess'. An article by one of *Science*'s own reporters said that Gallo's correction was likely to 'raise a few eyebrows. It could also have some legal ramifications.' Jim Swire, acting for the Pasteur Institute, would try to make sure of this. The mistake over the photograph came to the forefront of the legal battle. The hired gun of Townley and Updike had fired his first shot.

A central issue in the dispute between the Pasteur Institute and the US government was whether Gallo's researchers had used the French virus in order to help them to find the AIDS virus. Could it be that the sample of the French virus, LAV, somehow contaminated the virus discovered by Gallo, called

50

HTLV-3? Gallo has always denied this, and his researcher, Mika Popovic, stated that the culturing of HTLV-3 'was almost entirely confined to the tissue culture room 6B03A *where no LAV was ever used*' (Popovic's emphasis). (Contamination can and does occasionally occur in virology when a virus, say on the end of a pipette or circulating in droplets of water in the air, gets into scientists' culture media without their knowing.)

The French, however, were suspicious of Popovic's method of growing the virus. Popovic performed an unusual step in the isolation of HTLV-3. He pooled blood serum from ten AIDS patients and inoculated a particular strain of T-cells with this pool. Virologists usually take great care to keep their sera as pure as possible when they are trying to isolate a virus. Popovic nevertheless maintained that his method increased the chances of infecting the T-cells. Some virologists say this method was a stroke of genius on the part of Popovic – which ultimately proved to be successful.

Contamination is a dirty word to virologists. It continually threatens their research results, as well as their professional reputation. Gallo knows how embarrassing contamination can be. In 1975, he published a research paper announcing the discovery of a new human virus, which he called HL23. He suggested that this virus was involved with leukaemia. A year later, other researchers showed that the HL23 'virus' was in fact a cocktail of three ape viruses: gibbon-ape virus, simian sarcoma virus and baboon endogenous virus. HL23 was apparently the product of laboratory contamination. Gallo has described this example of contamination as 'bizarre' and has hinted that he was the victim of professional rivalry that led to sabotage. 'I mean, what could it be but sabotage? One contamination can occur, but three? In fifteen years I had had one contamination from a mouse. But three?' he later told a reporter for the *Washington Post*.

Other AIDS researchers have also had to confront the possibility of laboratory contamination. Max Essex and Phyllis Kanki, of the Harvard School of Public Health in Boston, announced in 1985 that they had found a quite different strain of the AIDS virus in people living in west Africa. The virus was so different that it warranted a new name, so they called it HTLV-4. When other researchers sequenced the genetic structure of this virus they found that it was practically identical to an AIDS-like virus that infects two species of monkeys, the rhesus macaque monkey and the African green monkey. These researchers suggested that either one virus could infect three different species, which is highly unusual, or that there had been a mix-up in the laboratory and one was being confused with another.

Certainly there had been the opportunity for a mix-up. Essex and Kanki had received samples of the monkey virus from the New England Regional Primate Research Center in Massachusetts at the end of 1984. The research centre had isolated the virus from macaque monkeys, which live in Asia and parts of Africa. But these monkey specialists believed that the animals had become infected with the virus by sharing cages with infected African green monkeys, which might have received the virus from their ancestors, who were infected when living in the wild. Captive macaques have over several decades often suffered from strange epidemics of AIDS-like illnesses. Scientists now think that this AIDS-like monkey virus was the culprit. The researchers from the primate centre believed that macaques in the wild are not infected with this virus. If the virus had been passed between caged monkeys, that could account for why the researchers found the same virus infecting two different species of monkey. But how can you account for the same virus infecting humans? It might be that humans in west Africa are infected with the identical virus that infects African green monkeys, or it might be that the monkey virus had contaminated the sample of HTLV-4 that Essex and Kanki had isolated.

Essex and Kanki have tried to pinpoint how the 'contamination' could have occurred, but they could not do so. Essex has said: 'There's no conventional cellular contaminant [i.e. a direct contamination with the virus], but we can't rule out an aerosol contaminant [i.e. when the virus is carried in water droplets in the air].'

52

One of the more confusing aspects of the AIDS story is the abundance of names that scientists have given to the 'AIDS virus'. Strictly speaking, the virus should not be called the AIDS virus. The virus does not directly cause the range of diseases seen in people who are diagnosed as having AIDS. Instead, the virus results in a breakdown in the immune system so that the body eventually becomes vulnerable to a whole range of 'opportunistic' infections.

Scientists, therefore, did not want to call the virus simply the 'AIDS virus'. The name that Gallo had originally given for the virus, human T-cell leukaemia virus type 3, or HTLV-3, eventually became human T-cell lymphotropic virus type 3. Gallo decided to change the name of his virus because it was clear that the AIDS virus did not cause leukaemia, like the other HTLVs. Nevertheless, it did have an affinity, or a tropism, for T-cells, hence the change to lymphotropic. And this still meant that Gallo could call the virus HTLV-3.

However, virologists were not at all sure that the AIDS virus should be classified alongside the HTLVs, which after all were cancer viruses. Retroviruses had three distinct classes: cancer viruses like the leukaemia viruses, a group called the foamy retroviruses and, finally, the lentiviruses, which included retroviruses that infected sheep, goats and horses. Where in all this did the AIDS virus fit in? Even when Gonda first took pictures of LAV/HTLV-3 he called it a lentivirus; evidently because under a microscope it looked so much like one.

Since Gallo announced his discovery of HTLV-3 in May 1984, the bulk of the press referred to the virus under this name. During 1985, however, virologists around the world were increasingly convinced that this was the wrong name for the AIDS virus. In May 1986, an international committee of virologists designed a more appropriate name for the virus. The committee called it the human immunodeficiency virus or HIV. From then on the virus that results in AIDS was called HIV. (The committee gave the AIDS-like virus that researchers had found in monkeys a similar name: simian immunodeficiency virus, or SIV.) Two members of the thirteen-strong committee disagreed, however. As far as Robert Gallo and Max Essex were concerned, the correct name for the virus was HTLV-3, and they both continued for some time to call the virus by this name.

Essex had found what he called HTLV-4 in west African patients at about the same time that Luc Montagnier, working with some Portuguese doctors, also found a virus that caused AIDS in west Africans. This virus appeared to be structurally quite different from his first virus. For instance, the antibodies of this new virus did not react strongly with the LAV/HTLV-3 virus, now called HIV-1. Montagnier therefore called this new virus HIV-2, denoting that it was a member of the same family of human immunodeficiency viruses, but that it was sufficiently different to warrant a distinct name.

When Montagnier and his colleagues were able to look at the detailed structure of this second AIDS virus they found something quite extraordinary: the first virus, HIV-1, shared just 42 per cent of its genetic structure with HIV-2. In other words the two viruses, which both caused AIDS in humans, were not very similar to each other at all. Clearly the two viruses had evolved from a common ancestor many years ago, possibly centuries. But now they are quite different, although they have a similar effect on the human immune system. When Montagnier compared HIV-2 with the monkey AIDS virus, SIV, he found that there was a stronger relationship between these two viruses than between HIV-1 and HIV-2 (although nothing like the 99 per cent relatedness of Essex's HTLV-4 to SIV). Montagnier proposed that all three viruses had shared a common ancestor in the past, but that HIV-2 and SIV had a more recent common ancestor (see figure 2).

How the HIVs in humans and the SIVs in monkeys relate to yet another type of AIDS-like virus, which this time infects cats, remains unresolved at the time of writing. Researchers at the University of California at Davis discovered this virus when a 'little old lady' walked into their laboratory one day claiming that some of her pet cats had AIDS. She said that she had noticed that some of them were suffering from very similar symptoms to the gay men with AIDS she had read

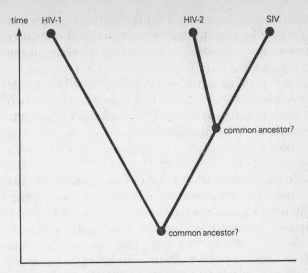

Figure 2. The second AIDS virus, HIV-2, is more closely related to the virus that causes AIDS in monkeys, SIV, than to HIV-1. SIV and HIV-2 may have shared a common ancestor years previously, just as HIV-1 and HIV-2 may have shared a common ancestor at an even earlier point.

about, including dementia. (She even claimed that her cats had picked up the diseases from a visiting cat from San Francisco, where, she had read, AIDS was rife.) Murray Gardner, a scientist from the university, told a meeting in Paris in October 1987 that some little old ladies 'are better at diagnosing dementia in their cats than in their husbands'. In the event, the lady's diagnosis was correct and Gardner announced in 1987 that he had found a virus that appeared to be the cause of the cats' malaise. Gardner and his colleagues found the virus in twenty-five unhealthy cats living in a colony of forty-three individuals. Only one of the eighteen healthy cats had antibodies to the virus. The diseased cats had a range of infections that suggested a breakdown in the immune systems of the animals. The researchers called the syndrome feline AIDS, or FAIDS. The virus is unlike cat leukaemia virus, which can cause the suppression of the immune system in cats, and unlike HIV. Although it seems likely that the human and the monkey immunodeficiency viruses shared common ancestors, the researchers were unsure how this cat AIDS virus related to the other viruses that cause AIDS in higher animals. In 1987, other researchers in the US even found a virus that appears to cause immune deficiency in cattle. It seems that the harder scientists look for viruses, the more they find.

Back on the twenty-sixth floor of the Chrysler Building in New York City, Jim Swire of Townley and Updike was contemplating a similar question concerning common ancestors to the two viruses discovered by Montagnier and Gallo. In May of 1986, the same month that the AIDS virus was officially christened HIV, Swire won a notable success in his bid to defend the Pasteur Institute's claim on the patent on the blood test for antibodies to the AIDS virus. The US Patent and Trademark Office, which had originally decided that the US government had the rights to the patent on a test developed by Gallo, now changed its mind: it decided to make the Pasteur Institute the 'senior party' in the patent claim. This meant that the US government, and Gallo, had to justify why they thought they should have rights to the patent in preference to the Pasteur Institute.

That month was not a happy one for Gallo. Virologists around the world began to call his virus by a name he did not like, and now the onus was on him to prove that he had the claim to the patent on his blood test. On top of this he had recently published an embarrassing correction to photographs his researchers had labelled wrongly, and he knew that other people, such as Jim Swire, were trying to make capital out of this error. Added to all this, the lawyers for the US government wanted detailed and time-consuming briefings from Gallo in order to prepare their case for what had now developed into a complex web of litigation. Gallo found that he was spending as much time on this as on his research.

There were in fact three simultaneous legal disputes between the Pasteur Institute and the US government. The first case concerned whether Gallo's laboratory had breached a contract with the Pasteur Institute not to use Montagnier's sample of LAV for commercial or industrial gain. The courts had originally decided that the letter limiting the use of LAV for research purposes only, and preventing its use for commercial gain, which Mika Popovic signed, did not constitute a 'government procurement contract', which had legal standing. The Pasteur later appealed against this ruling. The second dispute centred on the patent itself: who had the rights to royalties on a blood test for antibodies to the AIDS virus? And the third court dispute revolved around the US's law on freedom of information. The US government, the ultimate employer of Gallo, had already given Jim Swire several thousand documents relating to the discovery of HTLV-3. Swire complained, however, that many of these documents arrived in a random order – deliberate obfuscation, he claimed.

It was in relation to this last court case that Swire prepared to fire another salvo at the US government and the lawyers representing Gallo's side of the story. Swire had uncovered not one copy of the crucial letter from Matthew Gonda to Mika Popovic regarding the photographs of virus samples (see p. 49), but two. Both copies were identical save for one thing: on one version, someone had deleted the two items referring to LAV. Instead, the letter had a rather obvious blank space (see figure 3). Interestingly, on this version of the letter someone had scrawled a left-handed tick next to one of the blood samples that Gonda had analysed for evidence of virus particles. Swire had no idea who deleted this crucial information or why, but he suspected that someone was trying to conceal the fact that Gonda had taken good photographs of 'productive lentivirus' and that he had called it 'LAV'.

Swire prepared to present these two versions of the same letter to the US Federal Court in Washington DC as evidence that he could not rely on photocopies of documents from Gallo's laboratory. He wanted the originals. Swire was play-

NATIONAL
CANCER
INSTITUTE

FREDERICK CANCER
RESEARCH FACILITY

P.O. Box B Frederick Maryland 21701

December 14, 1983

Dr. Mika Papovic
Laboratory of Tumor Cell Biology
NCI-NIH
Bldg. 37, Room 6B22
Bethesda, Md. 20205

Dear Mika:

Enclosed are the results of all of the samples submitted by you and members
of your lab before 12-13-83.

		Virus	Comments
1)	PB Moweni	Negative	50% degenerated cells, lymphocyte series.
2)	BM Moweni	Negative	Lymphocytes.
3)	Lg. N. Moweni	Negative	Lymphocytes.
4)	PB Yau	Negative	Lymphocytes.
5)	BM Yau	Negative	Lymphocytes.
6)	HUT 78/LAV	Positive; Lentivirus	Productive lentivirus infection with all forms of virus maturation.
7)	T 17.4/LAV	Positive; Lentivirus	Lentivirus, same comments as #6 above.

ERSELL RICHARDSON (11-03-83)

		Virus	Comments
	11-1-83 *Mouse cell.*	Positive	Intercisternal A particles budding in endoplasmic reticulum. Many of the cells had degenerated.
2)	W6434 jk Da35 11-1-83	Negative	Lymphocytes.
3)	W6435 pw Da63 10-28-83	Negative	Lymphocytes.
4)	F-6357 Node Da28 10-28-83	Negative	Lymphocytes.

PRI   PROGRAM RESOURCES, INC. • Operations and Technical Support

Figure 3. The letter from Matthew Gonda to Mika Papovic: lawyers
for the Pasteur Institute claimed that the original (left) had been
tampered with (right) to conceal that the French virus, LAV, was
being grown.

ing poker with the lawyers acting for the US government. He
was trying to persuade them to settle the case out of court,
and needed the help of a little gentle persuasion. He thought
the two versions of the same letter would provide the necessary lubricant to settle the dispute once and for all.

Both sides were now edging towards a settlement out of
court. Very senior people in both the French and the American governments began to take a keen interest in the dispute.

NATIONAL
CANCER
INSTITUTE
FREDERICK CANCER
RESEARCH FACILITY
P.O. Box B, Frederick, Maryland 21701

December 14, 1983

Dr. Mika Papovic
Laboratory of Tumor Cell Biology
NCI-NIH
Bldg. 37, Room 6B22
Bethesda, Md. 20205

Dear Mika:

Enclosed are the results of all of the samples submitted by you and members
of your lab before 12-13-83.

			Virus	Comments
1)	PB	Moweni	Negative	50% degenerated cells, lymphocyte series.
2)	BM	Moweni	Negative	Lymphocytes.
3)	Lg. N.	Moweni	Negative	Lymphocytes.
4)	PB	Yau	Negative	Lymphocytes.
5)	BM	Yau	Negative	Lymphocytes.

ERSELL RICHARDSON (11-03-83)

		Virus	Comments
1)	11-1-83 + Mouse cells	Positive	Intercisternal A particles budding in endoplasmic reticulum. Many of the cells had degenerated.
2)	W6434 jk Da35 11-1-83	Negative	Lymphocytes.
3)	W6435 pw Da63 10-28-83	Negative	Lymphocytes.
4)	F-6367 Node Da28 10-28-83	Negative	Lymphocytes.

PRI PROGRAM RESOURCES, INC • Operations and Technical Support

The French government had become involved because it partly funded the Pasteur Institute. The French Minister of Health, Michèle Barzach, had written to her counterpart in the US, Otis Bowen, in the hope of putting pressure on both sides to settle a row that had now become publicly damaging. Scientists and governments were seen to be squabbling over patents when people were dying of AIDS. It was bad for science and bad for politics. As Swire once remarked: 'This was not your usual litigation. It would be settled, if at all, at the highest levels of government.'

Towards the end of 1986, lawyers for both sides began to draw up draft documents to prepare the way for a settlement. At the same time, Gallo and Montagnier began to work out an agreed chronology of the crucial discoveries. To help them to do this, Jonas Salk, a distinguished scientist who had himself been involved in an unseemly row thirty years previously with another scientist over the polio vaccine, offered his services as mediator and umpire. Salk knew how difficult a settlement would be: 'Something like this is like an illness in science, a psychosis. Something is out of order, people take sides. The patent issue set things off, but the coin of the realm was credit, not money.'

During 1986, Montagnier and Gallo were dogged by reporters wanting to know the background to their public row. *Nature* had obtained copies of laboratory notebooks and had, according to John Maddox, the journal's editor, 'embarked upon what is bound to be a complicated attempt to discover what the truth may be'.

Over Christmas 1986 and in the first few months of 1987, nearly half a dozen drafts of the settlement went back and forth between the scientists and lawyers of the two sides. During this period, however, *New Scientist* published a long article about the dispute, written by one of the authors of this book, which raised the temperature, according to Swire. For a short time the settlement appeared to be at risk. Both the Pasteur Institute and the US government agreed to put out an interim statement to the press condemning, without being specific, 'the inaccuracies which have appeared recently and in the past describing the dispute between the parties'. The statement added:

> The parties currently are negotiating an amicable settlement which will be honourable for all. This settlement would recognize the important contributions of Dr Gallo and his colleagues and Dr Montagnier and his colleagues leading to our understanding of AIDS and its diagnosis, and such a settlement should in no way be interpreted as providing either party an advantage over the other party.

Pressure now came from the highest levels in the French and American governments to settle the dispute in time for a planned visit by the Prime Minister of France, Jacques Chirac, to the White House. Gallo flew to Frankfurt in West Germany where he met Montagnier in a hotel room to put the final touches to a settlement. 'Monty brought a bottle of cognac,' Gallo later told the *Washington Post*, 'but I told him we wouldn't drink until we'd finished.' Gallo spent his fiftieth birthday with Montagnier to work out the final wording of a chronology of AIDS research. In the same week, a letter written by some of the most eminent scientists in their field, including several Nobel prize winners, such as Jonas Salk, appeared in *Nature*:

> We are pleased to note the approaching settlement between the United States government and the Pasteur Institute regarding patent rights related to the discovery of the human immunodeficiency virus (HIV), the virus of AIDS (acquired immune deficiency syndrome).

The letter ended:

> It is important to recognize that the discovery of HIV and its relation to AIDS is only the first step towards the ultimate conquest of this disease. We need to encourage our best scientists, both young and older, to engage in solving the urgent problem posed by the spread of this virus in the human population.

On the last day of March 1987, President Ronald Reagan and Prime Minister Jacques Chirac met in the East Room of the White House to announce to the world that the dispute between the Pasteur and the US government had been settled. President Reagan said that a new foundation would be established on part of the royalties from the blood tests developed by Gallo and Montagnier. The foundation would fund research into AIDS. 'This agreement,' Reagan said, 'opens a new era in Franco-American cooperation, allowing France and the United States to join their efforts to control this terrible disease in the hopes of speeding the development of a vaccine or cure.'

*

The French press called the settlement between Gallo and Montagnier 'the Yalta of AIDS'. The legal document itself was forty-three pages long and was signed by twenty people, including the US Secretary for Health and Human Services, Otis Bowen, the chairman of the Pasteur Institute and Nobel prize winner, François Jacob, a number of researchers in both laboratories, and, of course, Gallo and Montagnier. In addition, Gallo and Montagnier published in *Nature* a long 'official' chronology of the scientific events surrounding the discovery and research into the human immunodeficiency virus. For his role as intermediary in the writing of this summary, Gallo and Montagnier thanked Jonas Salk.

The terms of the settlement were that the names of both Gallo and Montagnier would appear on both of their patents on blood tests for antibodies to the human immunodeficiency virus. This would circumvent the tricky issue of who had the rights to such a patent in the first place. In addition, the settlement said that 80 per cent of the royalties that accrue from the patent, from 1 January 1987 to 27 May 2002, would go to the new research foundation. The object of the foundation, the legal agreement said, was this:

> The research foundation . . . shall support through grants the work of medical professionals and scientists throughout the world with respect to research into the cause, detection, prevention, treatment and cure of the disease AIDS and such other diseases caused by, or hypothesized to be caused by, a human retrovirus. The research foundation shall foster international cooperation and collaboration among medical professionals and scientists throughout the world with respect to the aforementioned research and shall also support through grants appropriate educational programmes.

Six trustees would sit on this foundation, three from the Pasteur Institute and three from the US Department of Health and Human Services. A committee of 'distinguished scientists' would assist these trustees in evaluating research proposals. A quarter of the money spent by the foundation would go towards research into AIDS in the developing world, particularly Africa.

The agreement also stipulated that each side should issue a statement to the press saying that they both disavow 'any statements, press releases, charges, allegations or other published or unpublished utterances that overtly or by inference indicated any improper, illegal, unethical or other such conduct or practice by any other Party or individual or their agents or employees.' Furthermore, clause 5 of the settlement, concerning the official history of AIDS research published in *Nature*, stipulated:

> The Parties hereto and those persons signing this Settlement Agreement in their individual capacities agree to be bound by such scientific history and further agree that they shall not make or publish any statements which would or could be construed as contradicting or compromising the integrity of the said scientific history.

In plain English this means that neither Gallo nor Montagnier, nor anyone in their laboratories, could comment further on the events leading up to the discovery and analysis of the AIDS virus. Clause 5 did not stop President Reagan commenting on the discovery of the virus two months later during his first speech on AIDS, at the Potomac Hotel in Washington DC. 'To think, we didn't even know we had a disease until June of 1981,' Reagan told his audience, which included Gallo and Montagnier. He continued, 'the AIDS virus itself was discovered in 1984'. Reagan's speechwriters had overlooked, perhaps intentionally, that Montagnier had discovered the virus in 1983.

Nature, which had in 1986 tried to uncover the facts behind the dispute, called for the burial of hatchets. Those who want to rake over the embers of the row, editor John Maddox wrote, could be likened to 'an army of ghouls'. Clause 5 in the legal document would make sure that neither Montagnier nor Gallo would encourage such ghoulish behavior. The blow-by-blow story is left for history to untangle – unless a Nobel prize committee does it first.

CHAPTER 5

TEST FOR INFECTION

Once scientists had found the cause of AIDS they could begin the task of designing a blood test to identify those people who might be carrying the virus. The theory is quite simple: find the virus, and then you can find the people infected with the virus. In practice, this is not quite so straightforward.

It was important to develop such a test because, by 1983, it became evident that the supply of blood to hospitals and clinics in the US had become seriously contaminated with the virus during the early 1980s. More and more people in the US who had received blood donations or products made from blood were developing AIDS. The blood supply in other countries, even if these countries did not have any cases of AIDS, had therefore also been put at risk. Many of these countries imported blood products, such as factor VIII, the blood-clotting protein, from the US. The international trade in blood and blood products had spread the virus far and wide. The trade had ensured that the virus lurking in the blood of unwitting donors had a passport to towns and cities around the globe.

Contamination of the blood supply had become a major political issue, especially in the US. When, in April 1984, Margaret Heckler, the US Secretary of the Department of Health and Human Services, hosted the press conference in

Washington DC announcing the discovery of the AIDS virus, she predicted that a test would be ready in six months and that it would be 100 per cent accurate. She was later proved wrong on both counts. The development of a commercial test took almost a year, and the test was not completely reliable.

From the outset, a 'blood test for AIDS' was fraught with technical and ethical difficulties. For a start the test was not a test for AIDS. It was not even a test for the virus. It was a test for antibodies to the virus and as such only indicated whether somebody had at some time in the past been infected, or 'exposed', to the virus. A positive result did not necessarily mean that the person was still infected with the virus, and it certainly did not automatically mean that they had AIDS, or would develop it. On top of this, the tests that scientists were to develop during 1984 were not totally accurate; there was always a risk that the test would wrongly diagnose people as having antibodies to the virus when they did not, and, conversely, wrongly diagnose people as not having antibodies when in reality they did.

Yet another problem with the blood test stemmed from its development as a means of screening donated blood to see whether AIDS antibodies were present. The test was made in order to stop infected people from donating blood, so clearing the blood banks of the virus. The use of the test as a diagnostic tool to tell patients their antibody status was in fact a spin-off of this. This immediately created difficult ethical dilemmas for the medical authorities: should they tell blood donors if the test indicated that they were carriers of the virus, and if so what sort of counselling should be given to them and their families or lovers? After all, many people would see a positive test as a death sentence. It could create all kinds of emotional and psychological problems. And what if the test was wrong? And what about third parties – employers, insurance companies and even medical staff looking after these people? Should they be told of positive results as well?

Such difficult questions had, however, become secondary to the task in hand – the blood supply must be cleared of the virus. The spectacle of children and babies developing AIDS as a result of contaminated blood transfusions, or blood products such as factor VIII, had resulted in hysterical articles in the press, describing these children as 'innocent victims', with the obvious insinuation that other 'AIDS' victims were in some way 'guilty'. The public's anxiety was aroused. No longer was AIDS a disease of gays and drug addicts. 'Ordinary' people, anyone who needed blood, were also now at risk of what was originally thought to be the problem of the minority.

In the US, this anxiety began to spread to the political establishment – 1984 was an election year for President Reagan, and AIDS, especially a blood supply that was contaminated with the AIDS virus, had threatened to become an important election issue. Heckler, and others within the Department of Health and Human Services, wanted a blood test quickly. Almost immediately after Robert Gallo had announced the discovery of the AIDS virus, 'HTLV-3' as he called it at the time, the health department launched its own campaign to develop a test that could detect antibodies to the virus. The department invited American drug companies to tender for licences to manufacture such a test, based on Gallo's HTLV-3 and the type of blood cell in which it appeared to thrive, the H9 line of white blood cells. Twenty companies took up the challenge. They knew that they could make millions of dollars from selling the test kits to blood banks and also to other organizations wishing to screen people for the virus. The US government was itself interested in screening people wanting to enter the military services. The potential world market for a 'test for AIDS' was enormous.

The US health department eventually gave licences to develop a blood test to five companies. These were Abbott Laboratories, from Chicago, Electro-Nucleonics from Maryland, Du Pont of Delaware collaborating with Biotech Research Laboratories of Maryland, Litton Bionetics also of Maryland, and Travenol Genentech Diagnostics of Cambridge, Massachusetts. The health department gave each company twenty-five litres of H9 cells infected with Gallo's HTLV-3, the AIDS virus, in June 1984. The race had begun to develop a test to rid the US blood supply of the virus.

At this time, the blood supply in the US and elsewhere remained open to contamination with the AIDS virus. As early as the spring of 1983, the medical authorities in the US had urged people not to give blood if they were in any of the high-risk groups for AIDS. European governments, including Britain and France, made similar requests. However, many people were afraid that closet gays and other people with a habit or lifestyle they had kept to themselves would continue to give blood because not to do so after being a regular donor might expose them in the eyes of their friends and families. Of course, blood banks would ask donors about their lifestyle in a confidential questionnaire, but this would not guarantee that those at risk of carrying the virus would refrain from giving blood.

In the period immediately after Gallo announced the discovery of HTLV-3, the medical authorities in the US were in a strange and difficult position. The cause of AIDS had apparently been found, and yet the authorities were seemingly helpless to prevent contamination of blood. Apart from urging high-risk groups not to donate blood, the only alternative precaution was a complete ban on the collection and sale of blood and blood products. It might be risky to receive donated blood, but it would be far riskier for most patients not to have blood transfusions at all.

There was in fact another suggestion, articulated by President Reagan in 1986: 'You know, there's a practical solution to that if someone would just announce it,' he told the Los Angeles Times, 'Why don't healthy and well people give blood for themselves? And then it can be kept in case they ever need a transfusion. They can get a transfusion of their own blood and they don't have to gamble . . .' This suggestion was not a new one. Doctors had considered autologous transfusions (i.e. of the patient's own blood) several years earlier, but had concluded that for the vast majority of patients it would be impractical and too difficult to organize.

Scientists, being the creatures that they are, saw a unique opportunity in the dilemma facing doctors in 1984. While they were waiting for the drug companies to develop blood tests for AIDS antibodies, they decided to conduct a study that would help them to understand more about how AIDS is transmitted. This study was to exploit the fact that some blood donors carrying the virus would inevitably give blood and this would end up being transfused into people who might then develop AIDS. Essentially the study was to see what would happen to people years after they received blood which was later found to have contained antibodies to the AIDS virus.

The director of the US's National Heart, Lung and Blood Institution, Claude Lenfant, explained the reasons behind the study to *Science*: 'Between now and the time a test is commercially available, we have a unique scientific opportunity to learn about the transmission of this disease.' He added: 'But it is important to emphasize that this study can only take place because the blood we want to collect and screen will be used in the usual course of blood banking now whether we do our study or not. An informed-consent form must clearly explain that we are not deliberately transmitting AIDS.'

It seems that even before a blood test had become available, ethical problems over an 'AIDS test' had arisen – informed consent became the issue. An informed-consent form is a document that people sign as evidence that they not only consent to a particular type of medical treatment, but that they are also aware of what is at stake – in other words, that they are fully informed of the implications of the treatment. This particular study planned to include 200,000 normal, healthy donors who did not fall into any of the risk categories for AIDS. They were asked to consent to a sample of their blood being stored until a blood test became available some months later. Their blood, meanwhile, would be used in medical treatment in the routine way. Once the test kits were ready, these stored blood samples would be tested for antibodies to the AIDS virus. The researchers would then follow up those people who had received blood from donors who had proved positive for AIDS antibodies.

At the time that this study was proposed, there was a great deal of debate within the American medical establishment about the ethics of 'informed consent'. According to *Science*, three out of four blood banks that were to take part in the study did not want to tell blood donors that their blood was to be tested for the antibodies. It was perhaps the first portent of the ethical debates to follow. As *Science* said:

> Because the first mass screening tests will be able only to detect antibody to AIDS in the blood, they will provide little clear information about whether the person is at risk of getting full-blown infection or whether he has simply been exposed to HTLV-3 and mounted a successful immune response. Heterosexual blood donors who test AIDS positive could be falsely labelled [as] homosexuals if the information leaked out. Healthy, non-promiscuous homosexuals also worry about the stigma that an AIDS-positive test would attach to them. All round, there is concern about what the information would mean to prospective employers or health-insurance companies . . . Thus, for many reasons, the likelihood that giving the donor complex, unclear, but frightening information will cause at least psychological stress is very high.

Another argument against telling a blood donor about a positive result was that this would attract to the blood banks the very people that blood banks did not want. Some homosexuals and drug users would actually donate blood in the hope that they would have a blood test for AIDS and so find out about their antibody status. The medical authorities in Britain, and later the US and elsewhere, quickly decided that any introduction of a test to screen blood for the AIDS virus at blood banks must not occur without offering, at the same time, anonymous testing at clinics for sexually transmitted diseases. In this way, people wanting to know whether they had antibodies to the AIDS virus could do so without giving blood. This evidently meant that a test that was being designed to screen donated blood was now also going to be used, for better or for worse, to 'tell' people whether or not they had been infected with the AIDS virus.

The implications of this took second place to clearing the blood supply of the virus. As *The New York Times* said:

'Blood has become the cornerstone of modern medicine, more significant to treatment than many drugs.' In the US alone, doctors administer about 12 million blood transfusions each year to about 3.5 million people. As a leading specialist in blood transfusions, Johanna Pindyck of the New York Blood Center, said in an interview with the same newspaper: 'It is a toss-up between transfusions and anaesthesia as to which has had a greater impact on surgery. You could put people to sleep and still not do the procedures that you are able to do now if it weren't for blood transfusions. Moreover the whole health care systems could not have developed without blood.'

The medical establishment clearly felt, therefore, that ridding the blood supply of the virus was a top priority. Scientists in the five companies in the US with licences to use Gallo's virus and the H9 cell line were working flat out in the summer and autumn of 1984 to develop test kits. At the same time, the American medical authorities were preparing advice for doctors, blood banks, and eventually the public in regard to the new test. The Department of Health and Human Services wrote to American physicians in February 1985 warning them that a blood test for antibodies to the AIDS virus would soon be licensed by the US Food and Drug Administration, the statutory body in the US concerned with the approval of new types of medications. 'This is NOT a test for AIDS', the letter emphasized.

The health department had decided that people should be made fully aware of what the 'AIDS test', as it was being referred to in the press, really was. Information about the significance of a positive result should be given to patients before the test takes place, the health department told American physicians: 'In addition, physicians and other health professionals should recognize the need for assuring confidentiality of test results because loss of employment or insurability may occur if positive test results become a part of the medical record,' said the department.

In its advice to the blood banks, the US health department recommended that each establishment should appoint a trained counsellor who would tell blood donors about positive results for blood tests when these occurred. The counsellors, the health department said, must 'understand the need for confidentiality and the severe psychological stress that reactive [positive] tests will cause in some individuals.' Even at this early stage in the application of a 'test for AIDS', the health authorities had a good idea of the demand that there would be from third parties to have access to the results of blood tests. 'In all cases,' the health department told blood banks, 'it will be extremely important to protect the confidentiality of donor identity in relation to test results and to restrict access to information . . . The misuse of such information could have serious consequences for both donors and blood establishments because positive test results could result in loss of employment or insurability.'

Having prepared doctors and blood banks for the impending blood test, the US Food and Drug Administration issued a licence for the sale of the first of the test kits in March 1985. The 'test for AIDS' had finally arrived.

Of course, the blood test was not a test for AIDS. It was a test for antibodies, the chemicals that the body produces when infected by the human immunodeficiency virus – HIV – which causes AIDS. From the outset it became obvious that detecting the presence of the virus itself was too difficult for a test that would have to be accurate, sensitive and fast. Trying to establish the presence of the enzyme reverse transcriptase, which is how the scientists at the Pasteur Institute in Paris first found the virus, in millions of blood samples was just not feasible. It would be far too cumbersome and time-consuming. Far better to design a test that would detect the presence of antibodies to the virus. After all, if a person had antibodies to HIV then this meant that they had been exposed to the virus and therefore could still be a carrier.

Trying to detect the virus directly has other problems. The amount of HIV circulating in the bloodstream of an infected person can vary depending on the stage of infection. Even when the amount of virus is at a maximum – perhaps at the stage when the virus has finished its incubation period and has begun to emerge from the T-cells – there is still precious little to detect accurately in any one blood sample. At this point in the life cycle of the virus the patient may in any case have begun to show many of the early symptoms of AIDS. The incubation period can last months or years and during this time infected people can pass the virus to others, although there may be little 'free virus' circulating in the bloodstream, posing enormous difficulties when it comes to detecting its presence. There have been more recent technical advances that permit scientists to detect the virus directly (see p. 79), but in 1984 researchers felt that the best blood test would be a test for detecting antibodies to the virus. Once the body produces antibodies to a foreign particle, such as a virus, they are relatively abundant in the bloodstream, and can almost always be found there long after the initial infection has taken place.

The American companies each working with twenty-five litres of HIV from Robert Gallo, courtesy of the US government, began to design a test based on a technique called enzyme-linked immunosorbent assay, or ELISA for short. This type of test for antibodies can exist in a number of different forms, but they all share the same basic technology. Most ELISA tests can be likened to layers in a sandwich. In the ELISA test for HIV, the first layer in the sandwich is viral proteins, which the companies purified as best they could and then attached to resin beads, or the sides of small depressions or 'wells' in a plastic dish. Scientists call these beads or wells the 'solid phase' and it means that anything sticking to this solid platform will not be washed away.

The next layer in the 'sandwich' consists of the antibodies to the virus. The virus is made of many different types of proteins (as we discuss in Chapter 7), and in the body each protein causes the production of a particular antibody that specifically recognizes and sticks to that protein. The companies attempted to use viral proteins that are particularly good at 'raising' antibodies. So, when these viral proteins – the antigens – are attached to the solid phase, adding a blood

72

sample containing antibodies to these proteins causes the second layer of the sandwich to form as the antibodies attach themselves to the antigens.

The problem now is how to discover whether or not there is a complex of antibody–antigen stuck to the solid phase; if the complex exists, then the blood sample has antibodies to HIV, and the test is positive. To do this, a further layer is added to the sandwich. This layer is another chemical that will bind to the antibody. It is in turn attached to a substance, called an enzyme, which turns the solution a different colour when a certain chemical is added.

An important feature of the ELISA test is that the solution is frequently washed so that anything not bound to the 'solid phase' (either directly or in a chemical complex) is washed away. If, therefore, there are no antibodies to the virus in the blood sample, then nothing, including the enzyme, will remain after washing. Therefore there will be no change in colour – a negative result (see figure 4).

There is a slight variant of this test which does not rely on an enzyme to change the colour of the solution. This technique uses instead the phenomenon called immunofluorescence, when certain chemicals show up or 'fluoresce' when illuminated with ultraviolet light. In this test the presence of antibodies to HIV is shown when the blood sample fluoresces in this way. The chemical that binds to the antibody to HIV and the 'sandwich' sitting on the solid phase contains a fluorescent dye instead of an enzyme. Fluorescence, then, rather than a change in colour, is the final visual signal indicating that antibodies to HIV are present in the blood sample. One of the first scientists to develop such a fluorescent test was Takeshi Kurimura of Tottori University in Japan. He had perfected his test by April 1985.

One of the first commercial tests to receive a licence from medical authorities, however, came from Abbott Laboratories and was based on the standard ELISA process. In a press statement in March 1985, Abbott claimed that its test identified forty-seven positive results in 18,000 samples of blood, 'indicating excellent specificity', the company said. (Speci-

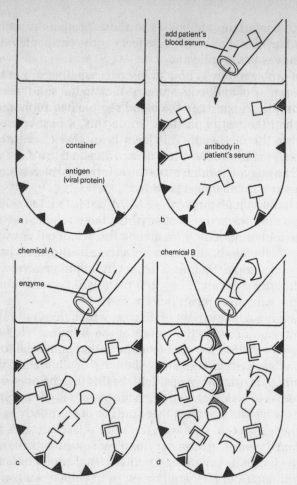

Figure 4. The standard blood test for infection with HIV is called the ELISA test. The manufacturers of this type of test supply specially prepared containers which have viral proteins, or antigens, stuck to the sides. These antigens cannot be washed off. Anything that sticks to the antigens cannot be washed away either. If antibodies to HIV are present in the blood sample added, they stick to the antigens. When chemical A is added, it will stick to these antibodies to HIV if they are present. Only if chemical A, which incorporates an enzyme, remains behind after washing, will chemical B change colour. So, with this type of test, a colour change indicates a positive result.

ficity, in terms of such blood tests, is defined as the ability of a test to determine healthy people as negative for the virus. In other words, a highly specific test gives rise to a low number of false positives.) Abbott failed to mention how many of these forty-seven positive results its test had *falsely* labelled as positive. A second, back-up test is needed to make sure that the positives are in fact true positives. There was to be a growing debate in the US, which Britain took so seriously that it delayed the introduction of its own blood test, about whether or not the new tests were throwing up too many 'false positives' – when the test wrongly decides that antibodies are present in a blood sample.

There are several reasons why a blood test can be inaccurate, and scientists have defined two terms that express these difficulties. The first is 'sensitivity', which is defined as the probability (usually expressed as a percentage) that the result of the test will be positive when the antibodies to HIV do in fact exist. A highly sensitive test, therefore, gives a low number of false negatives.

The opposite to sensitivity is 'specificity', which is defined as the probability (again expressed as a percentage) that the result of the test will be negative when the blood sample does not in fact contain HIV antibodies. In other words a highly specific test will result in a low number of false positives. Ideally a test should be 100 per cent specific – it should never identify the presence of HIV antibodies when they are not present – and 100 per cent sensitive – it should always identify the presence of HIV antibodies when they are in fact there.

In reality, the blood tests for HIV antibodies can only approach these optimum performances. The ELISA blood tests developed for the US in 1985 varied in specificity and sensitivity, but, in general, they had a sensitivity of about 97 per cent and a specificity of about 99 per cent. This means that a typical test would detect 97 out of 100 people infected with the virus, and identify as negative 99 people out of 100 who were not infected with the virus. Tiny changes in specificity can have dramatic consequences. For instance, if the true prevalence of the infection is one positive result in every

1,000 blood samples, a test which is 99.8 per cent specific will, from 1,000 samples, find one true positive and two false positives. If the test is only 99.2 per cent specific, however, it will detect one true positive and eight false positives – quite a dramatic difference.

At first glance, the relatively low sensitivity of the tests might seem a little surprising. It appears that the blood tests were missing, on average, three out of every hundred people infected with the virus. If millions of people are screened then this might mean that the test is failing to identify tens or hundreds of infected people, with catastrophic consequences for the safety of the blood supply. In fact this interpretation is wrong, because of another factor called the 'predictive value' of the blood test. The predictive value is a measure of how well the test performs in a given population and it can be applied to both a positive result and a negative result. For instance the predictive value of a positive result is the probability that the person will be infected if the result of the test is positive. And similarly the predictive value of a negative result is the probability that a person is not infected if the result of the test is negative.

Predictive values are not fixed; they change with the prevalence of the virus in the population. If a particular virus has infected 30 per cent of the population, for instance, then the predictive value of a positive result of a test with 97 per cent sensitivity and 99 per cent specificity is nearly 98 per cent (we will not go into the mathematics of how this was calculated). And the predictive value of a negative result with the same test is nearly 99 per cent, which is pretty good.

If, however, the virus in question is not very prevalent in the population being screened, which is the case with HIV in blood donors (it infects about 0.1 per cent of them), then the predictive value of a positive result changes dramatically – it is less than 10 per cent. Meanwhile, the predictive value of a negative result increases to almost 100 per cent.

This situation may seem odd at first, but common sense says that when a test result is negative for a virus that infects less than 1 in every 1,000 blood donors, it is more likely to be accurate than when the virus infects, say, twenty in a hundred people. This explains why the tests in question were better at identifying true negative results in a population of blood donors where the prevalence of the virus is extremely low, than in a population of homosexual men where the prevalence is quite high.

Because of these predictive values for a population of blood donors, it was not necessary to check a result that turned out to be negative. If a result proved positive, however, they had to be checked again, preferably using another test. If this too proved positive then a third type of test would be the final confirmation that the antibodies were indeed present.

The most common type of confirmatory test is known as the Western blot. This test can be a highly accurate with the correct reagents and experienced technicians. It is also, unfortunately, quite complex and therefore expensive to carry out. The Western blot works by first teasing apart all the different proteins in the virus. This is done by putting the proteins in a semi-fluid gel and exposing them to an electric field; the proteins, which are charged particles, move at different speeds, depending on their size. After a period of time the proteins are arranged in a column with, typically, the lighter proteins, having moved further, at one end and the heavier and larger ones at the other. The proteins are then transferred in these positions to strips of nitrocellulose gel.

It is then possible to treat the viral proteins attached to the nitrocellulose similarly to the 'solid phase' in ELISA tests. Antibodies, and radioactive chemicals which stick to these antibodies, will become bound to the nitrocellulose strips and the radioactivity will darken a photographic emulsion if this binding has taken place (see figure 5). The great advantage of Western blotting is that it identifies the *range* of antibodies that may be present in response to the spectrum of viral proteins of human immunodeficiency virus. Researchers have suggested that once a person becomes infected with HIV and 'seroconversion' (the production of antibodies against the proteins of HIV) occurs, antibodies first appear against the gp41 protein of the viral envelope and the p24 protein of the

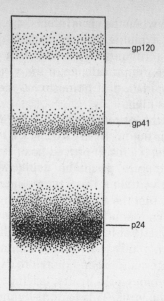

Figure 5. The pattern of a Western blot test shows whether a person is infected with HIV. If the test is positive, dark bands appear indicating the presence of antibodies to various viral proteins, such as p24, gp41 or gp120. The p24 band frequently shows the strongest reaction.

protein core of the virus (for a detailed discussion of the structure of HIV, see p. 95). Antibodies against the other proteins of HIV appear at a later stage.

Western blotting is not foolproof. False positives and false negatives do occur. However, the test looks for the presence of a range of antibodies and so, providing the laboratory technician looks for more than one antibody, the final outcome is highly likely to be an accurate result. A disadvantage of Western blots, which they do not in general share with other tests, is that the result relies on the skill of the technician interpreting the result. This is one of the reasons why Western blotting is carried out by research laboratories rather than commercial organizations, which may not have the skilled staff to make the correct interpretation of the visual pattern produced by the blot.

*

Technology is refining blood tests continually. Since the first tests became available in 1985, scientists have devised a number of different techniques to make the tests more specific and more sensitive. They have also developed tests that can detect viral proteins – the antigens – directly, rather than their antibodies. These tests rely on essentially the same principles as ELISA tests for antibodies, but this time antibodies from people already infected with HIV are used as the first layer of the 'sandwich', thus reversing the layers. The great advantage of these so-called antigen tests is that doctors can detect the presence of HIV before the person develops antibodies to the virus, in other words before seroconversion occurs. The problem, however, is collecting enough antibodies to HIV and purifying them in order to make these tests in bulk.

Another breakthrough in the development of tests to detect the presence of HIV is the result of the relatively new science of genetic engineering. These tests use synthetic viral antigens, made in the laboratory. The advantage of synthetic viral antigens over naturally occurring 'whole virus' is that researchers can pinpoint and use those particular viral proteins that are good at sticking to and therefore identifying a whole range of HIV strains. Since scientists found the second AIDS virus, HIV-2, they have been worried that tests would fail to detect both HIV-1 and HIV-2. At the time, most tests did in fact fail to do this. Tests that use synthetic viral antigens, however, can accurately detect both with a single test.

No doubt tests will become simpler and cheaper as time passes. In September 1987, the US Food and Drug Administration approved clinical trials of a new test that takes five minutes to give a result – whereas a typical ELISA test can take a couple of hours. A drop of blood from the person is placed on a plastic dish and a small latex ball loaded with synthetic viral antigens is added. If the blood contains antibodies, they can be seen with the naked eye forming clumps around the bead. This test could be freely available by 1988.

CHAPTER 6

A TESTING TIME

When blood tests became available in the US, in March 1985, the British government came under pressure to introduce tests too, using the American kits if necessary. The same happened in France. Neither country, however, introduced its testing programme until nearly six months after the US had begun its screening service. Coincidentally, both France and Britain were working on their own blood tests, and did not introduce screening programmes until these tests were ready. This brought criticism that the delay was unnecessary – both countries could have used the American tests and so have saved valuable time.

In Britain, the Department of Health and Social Security was particularly sensitive to the suggestion that it had delayed the introduction of a blood test in order to give a British pharmaceuticals company, Wellcome, the time to develop its own test. In France, the test the government approved was based on the virus isolated by Luc Montagnier and his colleagues at the Pasteur Institute in Paris, and subsequently used by Diagnostics Pasteur, a French company associated with the institute. It was easy to see why some people thought that chauvinistic pride in both countries had come before national priorities.

In the summer of 1985, under constant pressure to introduce a blood test, the British Minister of Health, Kenneth

Clarke, explained the reasons for not immediately using the tests that were already being used to screen blood in the US:

I understand and share the concern to get these tests in use as soon as possible. However, we must have tests which are accurate and can be trusted. A number of test kits are already available and in use abroad but reports from these countries suggest that the tests are not entirely reliable. We believe that no test should be introduced in the UK until its reliability has been established. There is no point in introducing a test which often fails to detect antibodies in blood or detects antibodies where there are none.

This was a slap in the face for the American health authorities, who had already approved blood tests in the US's screening programme. How accurate were these kits at that time? Nobody was more interested in this than the Americans themselves. In the first few months of the screening programme, the American Red Cross and other organizations concerned with collecting and distributing blood began to collate information on the number of positive results they found in the millions of blood donations they had tested. In one study, stretching from April to June 1985, and accounting for about 70 per cent of the American blood supplied during this period, researchers found that 0.85 per cent of the blood donations were positive for antibodies to HIV on the first test. Using a second test as a check, this was reduced to 0.25 per cent so-called 'repeatedly reactive' samples.

The prevalence of repeatedly reactive blood samples varied from one geographical region to another (0.14 per cent in the north-west to 0.29 per cent in the north-east of the US). Interestingly, repeatedly reactive rates were higher in females than in males – causing speculation that HIV was perhaps more prevalent in the heterosexual population than had been thought.

In fact these figures proved inaccurate. Many of these results were false positives and some researchers commented at the time that the blood tests then being used in the US 'cannot be used to define true positives'. Nevertheless, researchers were almost unanimous in saying that the tests had made the US's blood supply safer. Better to have a few false positives than to have no mechanism at all for detecting true positives.

There are several ways in which the result of a blood test can be false. Technicians can make mistakes such as diluting the reagents too much, splashing the specimens or flooding the wells so that the liquid from one well on the test plate runs into another. False negative results can occur simply because the person concerned has not yet developed antibodies to HIV even though infection has occurred.

But perhaps the most common reason for the high rates of false positive results detected in the early American tests was the design of these tests. The proteins of HIV were stuck to the solid phase, ready to capture any antibodies in the blood samples, as we have described. Unfortunately, as HIV buds from the human cell it infects, it becomes closely associated with human proteins in the membrane of the cell. These human proteins, called cellular antigens, also become attached to the solid phase of the test kit and so ordinary antibodies, which have nothing to do with recognizing HIV, can become attached to the solid phase – so producing a false positive result.

A way round this difficulty is to use what is called a 'competitive' test. Here, the HIV proteins are bound to the solid phase in the same way as in the non-competitive test, but the difference is that the blood specimen is added at the same time as a specific, pre-prepared antibody to HIV. This antibody is also linked to an enzyme that can turn the solution a different colour (see figure 6). In this way the real antibody to HIV (if it exists in the blood sample) and the manufactured antibody compete with each other for the limited 'landing sites on the solid phase. This means that a strong colour shows that there are no HIV antibodies in the blood sample, as all the landing sites have been occupied by the added HIV antibody, and a weak colour shows that the antibodies exist.

British researchers designed this type of 'competitive' test for HIV antibodies; a British pharmaceuticals company,

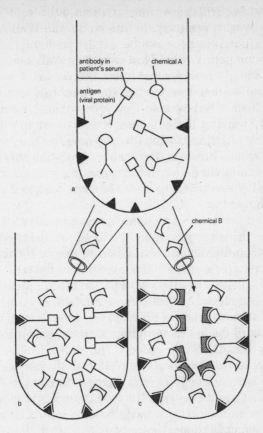

antibody in patient's serum

chemical A

antigen (viral protein)

chemical B

a

b

c

Figure 6. A competitive ELISA test is slightly different from the standard ELISA. Antibodies to HIV in the added blood sample compete for 'landing sites' (viral antigens on the surface of the container) with chemical A which incorporates an enzyme. If antibodies to HIV are present in the blood, few molecules of chemical A will have a chance to stick to the antigens. So chemical A will be washed away. Chemical B, however, changes colour only if chemical A is present. If chemical A has been washed away, no change in colour will occur: a positive result. If chemical A is present chemical B will change colour: a negative result.

Wellcome, developed the design into a product. The company, and the British government, praised the test for its low number of false positives. In one study, the Wellcome test produced nearly half the number of false positives that a similar, non-competitive blood test gave. The Wellcome test also had the added advantage that it required only one incubation period, rather than the two that the American tests needed. The Wellcome test was so innovative that it even won a Queen's Award for Technological Achievement in 1987, indicating that Britain was keen to promote the test to foreign buyers. Britain, however, was for a long time the only country to use a competitive test in the screening of donated blood. Eventually, researchers in West Germany designed a similar type of blood test.

The competitive test certainly proved to be valuable, but was the British government justified in delaying blood screening until August 1985, when in theory it could have introduced a programme at the same time as the US – in April 1985? The British business newspaper, the *Financial Times*, put it succinctly: 'The Wellcome Foundation . . . is likely to be the chief beneficiary of the relatively long time that Britain has taken to decide to introduce tests for screening blood supplies for AIDS.' It added: 'The delay in introducing screening in Britain has given Wellcome the chance to leap into the business of producing diagnostic kits for AIDS, a market that could be worth £100m–£200m worldwide by the late 1980s and which is now dominated by US companies, chiefly Abbott Laboratories.'

The blood transfusion service in Britain justified the delay on the grounds that British scientists needed to evaluate the tests with British blood. As John Barbara and Patricia Hewitt, of the National Blood Transfusion Service in north London, explained in a letter to *New Scientist*: 'It would have been irresponsible not to have seen for ourselves how the various tests performed in the hands of British transfusion microbiologists and when applied to British donors.' Furthermore, there was still the problem with the high rate of false positive results observed with some American tests. Tony Napier, the medical director of the blood transfusion service in Cardiff, pointed out that these tests would wrongly label many thousands of blood donors as being HIV positive, 'donors who will require interviews, repeat tests and sympathetic counselling. And for many of these, disruption of family and social life will be unavoidable . . . The current policy regarding the introduction of testing within Britain has not been a distant bureaucratic decision.'

Less than a year after the US had pioneered screening of donated blood, the American medical authorities were proclaiming a success. In Britain and France, and many other European countries, the transfusion services were also pleased with the outcome of screening. To the great relief of the British blood transfusion service, the testing seemed to confirm that its initial requests to high-risk people to stay away from blood banks seemed to have worked. Very few British blood donors were in fact positive for the virus.

In the US, figures of the number of people who had contracted AIDS from blood transfusions, but who had been infected before screening of blood had begun, showed that the risk was about 1 in 1,000. The percentage of AIDS cases in the US that had occurred as a result of blood transfusions had risen from about 1.2 per cent (12 people) in 1982, to a peak of 2.1 per cent (172 people) in 1985. Children with AIDS typically caught the virus from contaminated blood. By 1985, the percentage of children with AIDS who had contracted the disease as a result of blood transfusions was over 17 per cent (15 individuals) and still rose further in the following year as more children developed the symptoms of infections that had occurred before the screening had begun.

The situation in France was not quite as bad, but, nevertheless, the screening programme had, for the first time, given a clear indication of how prevalent the virus had become. The first major survey of donated blood, during the second half of 1985, revealed that there were nearly 1,000 HIV positive cases in nearly 1.5 million donations, which represented about 90 per cent of the country's supply of blood. This prevalence, approaching 1 in 1,000, worried the French medical authorities. As one scientist who took part in the study said, the measures taken by the French government to persuade high-risk groups not to donate blood 'were not fully efficient'.

In Britain, things proved to be a little better. The testing programme had shown that the virus was far less prevalent in those people giving blood in comparison to the US and France. The first big survey, between October 1985 and February 1986, on over a million donors throughout the country, showed that just nineteen were confirmed as being positive for the virus, a prevalence of about 1 in 55,000. The survey, however, revealed important and startling regional differences. In east Scotland, for instance, which includes the city of Edinburgh, the prevalence of the virus was as high as 1 in about 9,000 donors. A more detailed analysis of high-risk groups showed that about 65 per cent of intravenous drug users in the east of Scotland were antibody positive, whereas just 4.5 per cent of users were positive in the west of Scotland. Clearly there were cultural factors at work, the most likely being that Edinburgh drug users were more likely to share contaminated needles in the back-street 'shooting galleries' of the city than users in west Scotland.

Outside of the developed world, blood testing was far less methodical and frequent. There was another problem, especially in Africa. The tests themselves were more likely to be inaccurate, primarily because of the type of reagents in the test but also because there are more antibodies in the blood of people in Africa than in people living in more temperate climates. There are more diseases and infections in tropical and semi-tropical areas and so the immune systems of people living there have to cope with a greater number of foreign particles entering the body, thus producing more antibodies. Unfortunately, high level of antibodies in the blood can cause the HIV test to identify an antibody wrongly as being an antibody to HIV. Some scientists have described the blood of people living in Africa as 'sticky', meaning that there are many antibodies that can stick to a test kit and so give a false positive. Scientists have reported that the early HIV tests have, for instance, given positive results for people infected with the parasite that causes malaria, although these people are not in fact infected with HIV.

In one study of over 2,500 frozen blood samples that doctors had collected between 1981 and 1984 in five African countries, 9.3 per cent proved positive for HIV overall, and in one country the rate was over 23 per cent. A more detailed analysis of the tests, using the Western blot technique, however, showed that the vast majority of these were false positives. In this group of people, the scientists could confirm only two blood samples as being antibody positive. The researchers in this study, mainly from the Tropical Diseases Research Centre in Zambia, said that the results they had collected showed that, before 1984, the frequency of HIV in Africans was less than that of many European countries, and that 'the epidemic of AIDS started in central equatorial Africa at about the same time as the epidemic in north America'. A detailed analysis of the results of blood tests in Africa, therefore, seemed to dispel the popular notion that the virus had originated there and then spread to the West.

The statistics that began to emerge from Africa as a result of blood testing were not, however, conclusive — primarily because very few African countries could afford a screening programme. In fact, a survey by the World Health Organization in 1986 showed that just two out of forty-five African countries had begun routine screening for HIV antibodies. And yet there were many indications that it was in Africa, perhaps more than anywhere else in the world, that the problem of AIDS had become most acute.

There is one group of people for whom AIDS had become a cruel irony — haemophiliacs. AIDS struck the lives of these people and their families just ten years after improvements in medical technology had at last allowed many haemophiliacs to lead almost normal lives. By the early 1970s, people suffering from even the severest forms of haemophilia

87

could for the first time avoid the once-regular visits to hospital by injecting themselves at home with regular doses of factor VIII, the vital blood-clotting agent collected from blood donors. A decade later, it became apparent that the same batches of life-preserving protein were the cause of an infection with a life-threatening virus.

Long before AIDS was even defined, medical scientists knew that treating haemophiliacs with batches of factor VIII made from other people's blood had its risks. The most common problem was, and still is, the contamination of the blood product with the viruses that cause the various types of hepatitis, such as hepatitis-B and non-A-non-B hepatitis, which can lead to severe liver disease, including cancer. The problem with factor VIII is that it is made from up to 30,000 separate donations of blood. In fact, pharmaceuticals companies make factor VIII from blood plasma, the clear liquid in which the red blood cells float. People can donate their blood plasma more frequently than whole blood because it takes longer for the body to replace red blood cells than it does to replace plasma. In the US, people have made money by selling their plasma to pharmaceuticals companies, a trade that many other countries do not permit.

During the manufacture of factor VIII, companies freeze the blood plasma so that the blood proteins, including factor VIII, become concentrated in a 'cryoprecipitate'. Unfortunately, any virus in the pool of plasma also becomes concentrated in the same precipitate. It is very difficult to separate these viruses from the factor VIII. When haemophiliacs began to show the symptoms of AIDS, it confirmed many scientists' early suspicions that a virus was the cause of the problem. The warnings to the community of haemophiliacs, however, came too late for the majority. The virus had already infected many haemophiliacs as a result of the repeated injections of factor VIII that these people have to live with.

In many countries, such as Britain, France, West Germany and Australia, the proportion of people suffering from severe haemophilia who had become infected with HIV before precautions were taken had risen to well over 50 per cent by 1987.

In the US, the proportion approached 100 per cent. Clearly the situation was much worse in the US, and all the evidence suggested that factor VIII imported into European countries from the US was much more likely to be contaminated with HIV than factor VIII made in Europe. One suggestion was that some American companies who paid 'donors' for blood plasma had attracted the wrong clientele because of the financial inducements. The argument was that these donors would be more likely to be in high-risk groups, such as drug users, if they were so poor as to have to sell their plasma, and such people were less likely to be honest about high-risk activities.

In many countries, including Britain, France and Japan, there was great opposition to importing factor VIII from American sources. In Britain, for instance, the government came under intense pressure to uphold a ten-year-old promise that the country would become self-sufficient in blood products such as factor VIII. In 1987, however, the promise had still not been fulfilled. According to the Haemophilia Society, a British charity representing 5,000 haemophiliacs, Britain was only making a fifth of the nation's total demand for factor VIII in 1987: 'If the UK had processed sufficient voluntary donated plasma into factor VIII,' the society said, 'the number of people infected would be substantially less because the use of heavily contaminated material from abroad would have been avoided.'

In 1984, scientists discovered that heating factor VIII to high enough temperatures could kill certain viruses, such as hepatitis viruses, in the same way that pasteurizing milk killed many of the bacteria in milk. By 1985 most companies and laboratories making factor VIII had begun to heat it during the manufacturing process to kill HIV. A year later, however, some doctors treating haemophiliacs began to have suspicions about whether heat treatment actually worked. A handful of haemophiliacs had apparently developed antibodies to HIV a year after being given the heat-treated factor VIII. The question was whether they had become infected with virus that had survived the treatment.

The company at the centre of the controversy, Armour Pharmaceuticals of the US, decided to recall certain batches of its product from the market. The British government later withdrew Armour's licence to sell factor VIII in the UK because of the fears associated with its product. Armour's manufacturing process involved heating factor VIII to 60°C for 30 minutes. The Blood Products Laboratory, a government establishment at Elstree in Hertfordshire, recommended that factor VIII should be heated to 60°C for 72 hours. The disadvantage of higher temperatures for longer periods, however, is that more factor VIII is destroyed in the process and so the product becomes more expensive to make, although safer.

Armour decided to develop a method of cleaning up factor VIII that would produce a much purer product, free of virus. The method, developed by researchers in California in the early 1980s, was to filter the factor VIII through a column of antibodies. These would behave like tiny magnets pulling out molecules of factor VIII and letting anything else pass through. In this way, the company believed, the final product would be more than 99 per cent pure, and would not even need to be heat treated, although this could be done for extra safety. The company received for a licence to sell the product in the US in 1987.

The ultimate breakthrough for haemophiliacs, however, will be the manufacture of factor VIII by the new science of genetic engineering. The idea is to insert a gene that can make factor VIII into a microorganism, which will then produce the human blood protein in large quantities for harvesting. There is even one plan to insert the human factor VIII gene into cows so that the protein can be collected in the animals' milk.

Genetically engineered factor VIII and heat-treated factor VIII have come too late for the thousands of haemophiliacs throughout the world who have become infected with contaminated blood products. Many hundreds of these have already developed AIDS, and some have died. Dozens of wives and lovers of haemophiliac men have already learnt that they have also become infected with the virus. They, like their antibody-positive partners, live in the knowledge that

AIDS is a real possibility. The development of a blood test has been a blessing to many who would have otherwise received contaminated blood products or blood transfusions, and who would have possibly infected their sexual partners as well, but to many others the test has only brought the frightening prospect of premature death.

CHAPTER 7

ANATOMY OF A VIRUS

The efforts of researchers studying AIDS and the human immunodeficiency virus are unprecedented. Never in the history of medicine have scientists found out so much about a disease in such a short time. In 1982, medical researchers tentatively suggested that the cause of AIDS might be a virus, perhaps a new one. At that stage, they did not know that thousands of people without symptoms had already been exposed to this novel infectious agent. Now, they probably know more about the human immunodeficiency virus than about any other virus.

One of the most fundamental characteristics of HIV is the unusual way in which it stores its genetic information. This is in the form of ribonucleic acid (RNA). Most living organisms, by contrast, have genetic material in the form of deoxyribonucleic acid (DNA). The DNA is the blueprint of life: it carries the genetic information unique to the organism. This information is held in the form of the particular sequence of the small molecules that make up the long strand of DNA – equivalent, in the analogy used on p. 43, to the beads of a necklace. From this sequence of molecules, the cells of plants and animals make a mirror-image or complementary version of the genetic material. This copy is called RNA (ribonucleic acid). The RNA contains the information that the cell needs in order to make proteins.

Proteins are a vital component of living organisms. These molecules are made up of strings of smaller molecules, called amino acids. These strings are often folded, giving the protein a specific shape which reflects the molecule's function. Proteins come in a huge variety. Some are structural and highly specialized, such as those that make up human hair and nails. Other proteins may be involved in the daily life of a cell, making sure that it responds to signals, or receives enough food and oxygen, or eliminates waste materials. Some of these proteins will be of a type called enzymes. Enzymes play an important role in making sure that certain chemical reactions in the body take place. Enzymes in the human gut, for example, break down food into a form that the body can absorb.

Some organisms are simpler than others. Perhaps the simplest of all are viruses. These tiny particles have evolved a method of hijacking the cellular machinery of other organisms for making proteins. A virus is not a complete cell: it has only some genetic material and a protective 'coat' of protein. It is not alive in the strict sense of the word, for it has no means of reproducing itself – until it enters a cell of its host.

Some viruses contain DNA as their genetic material. Once the virus has attached itself to the host cell and inserted its DNA, the cell is deceived into making viral proteins. Some of these proteins are the enzymes necessary for the synthesis of more viral DNA. Many new viruses assemble themselves from these component parts, bursting free from the host cell, which may die in the process.

Other viruses, including HIV, contain RNA instead of DNA. As explained earlier, HIV belongs to a family of viruses named the retroviruses. 'Retro' means backwards, and retroviruses are so-called because the virus persuades the host cell to convert viral RNA back into DNA, contrary to the cell's normal method of operation, which involves making RNA from DNA. When HIV infects a cell, its outer envelope fuses with the membrane of the cell (see figure 7). This releases viral RNA into the cell, along with an enzyme which tells the cell to make DNA from the RNA. This enzyme, as we said

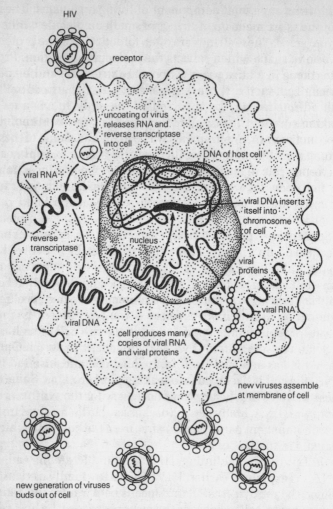

HIV

receptor

uncoating of virus
releases RNA and
reverse transcriptase
into cell

DNA of host cell

viral RNA

viral DNA inserts
itself into
chromosome
of cell

reverse
transcriptase

nucleus

viral
proteins

viral RNA

viral DNA

cell produces many
copies of viral RNA
and viral proteins

new viruses assemble
at membrane of cell

new generation of viruses
buds out of cell

Figure 7. *The life cycle of the human immunodeficiency virus. It takes over the cell's machinery for making proteins, channelling the activity of the cell into the manufacture of many new viruses.*

in Chapter 3, is called reverse transcriptase. It is unique to the virus and does not occur in human cells. Such individuality may be important when it comes to designing drugs to combat

HIV infection, for the trick is to find some way of knocking out the virus without destroying the human cells in which it lives.

The viral DNA then enters the nucleus and integrates itself into the DNA of the cell. Once there, the viral DNA lies dormant. This is the latent stage of infection, the incubation period. It can last for months or years. Eventually, when some trigger activates the cell, the viral DNA starts to direct the production of viral components. There are many theories about what the trigger might be. For example, there is some evidence that subsequent infections with other viruses, such as herpes viruses, can activate infected cells. Whatever the trigger, the result is the manufacture of viral protein and viral RNA – the two main components of HIV. The viral proteins migrate to the surface of the host cell, where they stick out through its outer membrane. The remaining elements of the virus, including the RNA, also assemble themselves beneath the cell membrane. Then, by a process known as budding, multitudes of new viruses detach themselves from the host cell, borne away in the bloodstream to attack other cells. Each virus has a diameter of only 0.1 micrometres. A cube containing a thousand million viruses would measure just one tenth of a millimetre across – barely visible to the naked eye.

During budding, the virus takes part of the cell's outer fatty membrane with it. Molecules of protein sit at regular intervals in this viral membrane or envelope (see figure 8). These proteins are called glycoproteins because they have molecules of sugar attached to them. HIV has two glycoproteins, called gp120 and gp41. The 'gp' stands for glycoprotein; the numbers reflect the sizes of the molecules. The viral membrane also bears proteins called HLA, or 'cellular' antigens, which are derived from the membrane of the host cell.

Inside the viral envelope is a structure called the core shell. This layer may be made of a protein known as p18. Researchers in California believe that the core shell has a faceted appearance. The core shell conceals the core itself, which contains the genetic material of HIV, the RNA. The core, which may be made of a protein called p24, seems to

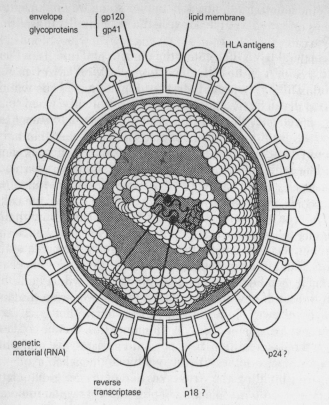

envelope glycoproteins — gp120 / gp41

lipid membrane

HLA antigens

genetic material (RNA)

reverse transcriptase

p18 ?

p24 ?

Figure 8. A model of the structure of the virus. The virus derives its lipid membrane from the human cell by a process of budding.

have a highly organized structure too, being in the form of a hollow cone. Studies with powerful electron microscopes suggest that the narrow end of this cone is open, while the wider, closed end is dimpled, rather like the base of a champagne bottle. Scientists believe that the enzyme reverse transcriptase, so vital to the virus's success in attacking cells, is somehow associated with the RNA in the core of the virus.

The virus attacks human cells with the help of the glycoproteins in its envelope. These carry a special binding site that recognizes that and attaches to another type of glycoprotein, a molecule called CD4, which is found on the membranes of some human cells. One of the commonest kinds of cell to carry the CD4 molecule is the type of white blood cell called the T-helper cell. This affinity to T-helper cells, which play a central role in the immune system, is the key to the paradox of HIV infection. HIV attacks the very cells which should normally be able to eliminate it.

The function of a person's immune system is to recognize and eliminate foreign substances, called antigens, that have entered the body. Any foreign material can act as an antigen, whether it is the protein coat of a bacterium or a virus, a cancer cell, or a transplanted organ. Every individual encounters scores of antigens every day. Most of the time, the cells and molecules that police the body for antigens deal with the intruder. At other times, perhaps because the body has never encountered that antigen before, the person succumbs to the infection. Eventually the body mounts an immune response and manages to fight off the bacterium or virus concerned. A key element in this response is the production of antibodies, protein molecules that recognize and bind to specific antigens and help to eliminate these foreign particles.

Complex interactions between many types of cells and molecules – not just antibodies – regulate the immune response. Some of the cells have the task of engulfing foreign antigens. Macrophages, for example, are large cells that can swallow and destroy bacteria and, sometimes, viruses. Other cells that take a key part in the immune response are the lymphocytes – white blood cells. There are two types of lymphocyte, T-cells and B-cells. When a B-cell encounters an antigen that it recognizes, it becomes transformed into a plasma cell (see figure 9). Plasma cells manufacture quantities of antibodies that recognize the original antigen. The antibodies bind to the antigen, inactivating it in a variety of ways. Antibodies that bind to bacteria, for example, may make it easier for macrophages to kill these microorganisms.

Viruses, unlike bacteria, are not alive in the strict sense of the word because they have no means of reproducing until they invade a host cell. So scientists talk about neutralizing

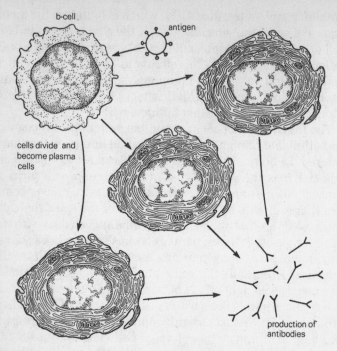

b-cell

antigen

cells divide and
become plasma
cells

production of
antibodies

Figure 9. Antigens stimulate B-cells to divide and mature into specialized cells called plasma cells. These cells manufacture antibodies
that recognize the original antigen. In people with AIDS, some B-
cells produce antigens at the wrong times while others fail to
respond to new antigens when they should.

viruses, rather than killing them. In many cases, antibodies
can neutralize viruses, perhaps by binding to the virus at the
point where the virus normally attacks the cell. The body can
then eliminate the virus. Antibodies that are capable of this
feat are called 'neutralizing antibodies'. Unfortunately for
people infected with HIV, although the body produces some
neutralizing antibodies against this virus, they seem to be
unable to eliminate the infection. At the end of 1987, scientists
did not know why this is so, or what the role of these antibodies is in preventing the disease from progressing.

The other class of lymphocyte is the T-cell. T-cells regulate

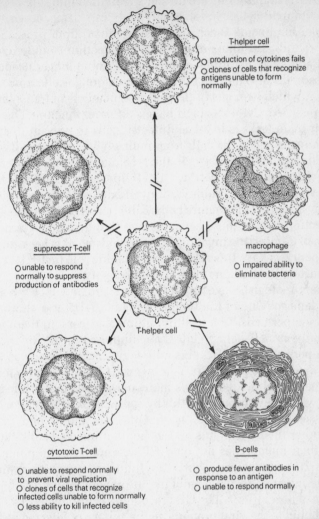

T-helper cell
○ production of cytokines fails
○ clones of cells that recognize
antigens unable to form
normally

suppressor T-cell
○ unable to respond
normally to suppress
production of antibodies

macrophage
○ impaired ability to
eliminate bacteria

T-helper cell

cytotoxic T-cell
○ unable to respond normally
to prevent viral replication
○ clones of cells that recognize
infected cells unable to form normally
○ less ability to kill infected cells

B-cells
○ produce fewer antibodies in
response to an antigen
○ unable to respond normally

Figure 10. The T-cell plays a central role in the immune system.

the immune response by either suppressing or encouraging
other components of the immune system (see figure 10). There
are two kinds of T-cell. The first type has a molecule called
CD8 on its membrane; these cells used to be called T8-cells.

The cells in this group are further subdivided into suppressor or cytotoxic cells. Suppressor T-cells can damp down the response of some B-cells, preventing them from producing their antibodies indefinitely once an infection is under control and such quantities of antibodies are no longer needed. These cells also initiate the rejection of foreign cells from the body, which happens in the rejection of an organ transplant, for instance. Cytotoxic cells have a different function. One of their main tasks is to attack infected cells to prevent microorganisms inside the cells from multiplying and spreading.

The second type of T-cell, the type that HIV infects, is the T-helper cell, which carries CD4. Helper cells coordinate the activity of B-cells, suppressor/cytotoxic and macrophages. They also direct immature cells, called monocytes, to sites of infection, where these cells develop into macrophages capable of eliminating invading microorganisms. It is because T-helper cells have this central coordinating role that HIV can devastate the immune system so effectively.

Immunologists have begun to understand how HIV wreaks its damage. One of the typical signs of AIDS is a shortage of T-helper cells. In a healthy, uninfected person, there are usually more T-helper cells than suppressor/cytotoxic cells. In a person with AIDS, this ratio is reversed. The reason for the drop in the number of T-helper cells is not entirely clear. Probably, HIV directly kills many of the cells that it infects. The virus may accomplish this by overwhelming the cell with masses of viral protein, so that the cell bursts and dies. Another theory depends on the fact that infected cells carry on their membranes not only CD4 receptors – the proteins that HIV recognizes – but also viral envelope proteins that have migrated to the membrane in preparation for the budding of new viruses. The CD4 receptors on uninfected T-cells will bind to the viral proteins on the surface of infected cells. So if two infected cells meet, or an infected cell comes into contact with an uninfected T-helper cell, the two cells can bind together. Adjacent parts of the membrane of an infected cell can also become bound to each other, disrupting the normal configuration and activity of the cell.

Scientists frequently observe this kind of interaction when they grow batches of infected cells in the laboratory. The result is that many cells merge, forming large structures with many nuclei, called syncytia, which do not survive for long. It is not always possible to establish what happens in the body from laboratory observations such as these. It may be that these events do not occur in the body. However, scientists have seen large cells with many nuclei in some tissues from patients with AIDS. This suggests that a similar process may account for the loss of some T-helper cells in the body.

A further explanation for the dearth of T-helper cells may be that because such infected cells display viral proteins on their surface, the body's own cytotoxic T-cells recognize them as foreign, and go about eliminating them.

The low number of T-helper cells is not the whole story, however. Even in people with AIDS who have very low numbers of T-helper cells, only a very small proportion of those cells appear to be infected at any one time. Yet infection with HIV seems to have wide-ranging effects throughout the immune system. For a start, T-helper cells from patients with AIDS fail to function normally. Some researchers have shown that such cells cannot respond normally to a stimulus from an antigen which would usually cause them to divide and proliferate. Possibly, viral proteins floating free in the blood bind to the CD4 receptors of uninfected cells. Once these receptors are blocked, these cells are unable to respond appropriately to antigens and cannot stimulate other cells of the immune system.

Some of these abnormalities probably occur because HIV can also infect other cells of the immune system. Although T-helper cells are the cells most severely affected by HIV, other cells, such as macrophages, also carry the CD4 receptor, which makes them susceptible to the virus too. Macrophages, the cells that engulf cellular debris and other foreign matter, start their life as smaller, less specialized cells called monocytes. Both monocytes and macrophages have some CD4 molecules on their surfaces, though these are more sparsely distributed on macrophages than on the membranes of T-

101

helper cells. The virus seems to be less effective in killing monocytes and macrophages. As a result, these cells probably form a reservoir of infection. They may also be the cells that carry the infection to the brain.

Some immunologists believe that the main defect in the immune system in people suffering from AIDS may lie not with the T-helper cells but with macrophages. Macrophages and related cells have an important role in presenting antigens to T-helper cells in order to activate them, and it may be abnormalities in this function that account for many of the immunological aberrations common to people with AIDS.

Still other abnormalities may occur because HIV interferes with the normal release of chemical signals by cells of the immune system. Some of these molecules are called cytokines, substances that stimulate cells to grow and become specialized for a particular function. One cytokine is called interleukin-2. T-cells normally produce interleukin-2 when stimulated by certain antigens; this cytokine causes other T-cells to divide and mature. T-cells that are infected with HIV, however, produce abnormally low amounts of interleukin-2 when stimulated. Researchers have also found that when they add other cytokines to infected cells, these stimulate the virus in the cells to begin replicating. Puzzled by these observations, researchers are still trying to determine the role of these chemicals in people infected with the virus.

CHAPTER 8

IN THE GRIP OF THE VIRUS

Whatever the exact causes of the failure of the immune system in people with AIDS, the effect is the same. A person infected with the virus develops severe immunodeficiency and the body is no longer able to fight off infections and malignant cells. Various viruses and bacteria seize their opportunity and multiply while the body's defences are down. The defects in immunity also allow certain cancers to develop. People with AIDS are prone to a type of skin cancer called Kaposi's sarcoma, and to another kind of cancer called lymphoma, which develops in the lymph glands.

In the few years since doctors first described the syndrome of infections and tumours known as AIDS, researchers have been hard at work untangling the new evidence about the course of the disease. During 1986 and 1987, medical scientists established a great deal about how the body responds to infection with HIV. Much of this information has become available as a result of the development of new blood tests.

As already mentioned, the first tests capable of telling whether someone was infected with HIV relied on the detection of antibodies. A new generation of tests, however, allows researchers to detect viral antigen – in other words, viral proteins – rather than the antibodies that the body makes in response to these proteins. One surprising observation that these tests have made possible is that antibodies to the virus may take several months to appear.

Researchers know, from studying the handful of cases where health-care workers have become infected after stabbing themselves with a needle contaminated with blood from infected patients, that antibodies generally appear a few weeks to a few months after exposure to the virus. Yet research published in 1987 by Finnish and American scientists, led by Annamari Ranki, suggested that some people who have been exposed to the virus may take over a year to develop detectable antibodies. Their results imply that some people who had negative results to conventional tests for antibodies as long as, say, three months after exposure to the virus, may have been falsely reassured that they were not infected. Ranki and her colleagues studied nine people with HIV infection. The researchers had at their disposal stored samples of serum from these patients, taken before they developed antibodies. The second part of the study involved twenty-five males, the sexual partners of infected men. The twenty-five were apparently not infected. According to the standard screening test, ELISA (see p. 72), they had no antibodies to the virus. The researchers looked for other signs of the virus, such as viral antigen and viral genetic material, in the serum from both groups of men. They also studied the development of antibodies with the sensitive Western blot test.

In the samples from the nine patients, the researchers found evidence of viral proteins and/or low levels of antibodies as much as six to fourteen months before the conventional ELISA test could detect antibodies. Of the twenty-five apparently uninfected men, five had similar evidence of infection, which had appeared between sixteen and thirty-four months before the development of antibodies detectable by the ELISA test.

Interestingly, although all the people with such evidence of latent infection had typical numbers of T-helper cells, in half of them these cells failed to respond normally to an antigen which would usually stimulate them to divide. This suggests that the virus was already having some adverse effects on the immune systems of these people.

Some of the same scientists have also found that in people with such latent infection, a full antibody response occurred only after the individuals had become infected with other viruses such as cytomegalovirus, Epstein-Barr virus (which causes glandular fever, also known as mononucleosis) or hepatitis-B virus. These viruses all contain DNA, and the researchers suggested that they might play a role in enhancing the replication of HIV. The patients in this study all started to suffer falling numbers of T-helper cells following the appearance of antibodies. Exactly how these intriguing studies tie in with the observations that antibodies often develop within two weeks to three months following exposure is not clear. One explanation may be that, while some people succumb quickly to the infection, others manage to fight it off for some time. Alternatively, the length of the period between exposure and the appearance of antibodies to the virus may depend on the route of exposure. Someone who has stabbed themselves with a contaminated hypodermic needle may react differently to someone who has contracted the virus as a result of having sexual intercourse with an infected person over an extended period.

Many people report having a minor illness shortly after their first exposure to the virus. This illness is similar to many other viral diseases, such as influenza, glandular fever or rubella (German measles). People feel tired and unwell, with aching muscles and joints. They may have a sore throat, swollen lymph glands, a high temperature, a rash or diarrhoea. Symptoms that indicate that the nervous system is involved may appear: the person may become confused and disorientated, with loss of memory or changes in personality. These symptoms last for about eight to twelve days, and the person usually recovers without the need for medical attention.

Tests carried out at this time show that viral antigens, particularly the core protein called p24, are present in the blood. Yet tests for antibodies usually prove negative at this stage. Antibodies often take two weeks to three months to appear. At around the time that the antibodies develop, the core protein of the virus, the p24 antigen, vanishes. This disappearance may be the result of a temporary immune response. A period of latent infection then begins, during which most people remain well. This period can last for months or years. Yet a proportion of people will at some point experience symptoms caused by the infection.

There are several patterns of illness caused by infection with HIV, some of which are eventually classified as AIDS if the symptoms become severe enough. Many people infected with the virus develop persistently enlarged lymph nodes. This condition is called persistent generalized lymphadenopathy (PGL) or lymphadenopathy syndrome (LAS). In many cases, the development of PGL is what prompts the person to visit a doctor. The enlarged lymph nodes appear most commonly in the neck and the armpit and under the jaw. (Enlarged nodes in the groin do not count in the definition of this syndrome, because such swelling commonly occurs as a result of other sexually transmitted infections.) The affected nodes are symmetrical – for example, in both armpits – and at least one centimetre in diameter. They do not feel tender.

Some people progress to PGL early on in the course of infection while in others this is a late development. Doctors have carried out many studies to try to determine the significance of PGL and whether this condition makes it more or less likely that the person concerned will develop AIDS. So far, research suggests that between about 10 and 30 per cent of people with PGL progress to AIDS, but the proportion goes up as time goes by. Scientists at the Centers for Disease Control in the US, for example, found that just under 30 per cent of a group of seventy-five infected men with swollen lymph glands developed AIDS over a five-year period. The risk of developing AIDS did not appear to go down during these five years. Whether someone develops AIDS probably depends on how long that person has been infected with the virus.

Another team of researchers has studied over 1,000 individuals in San Francisco, as part of the San Francisco Men's Health Study. Out of the 1,000 men, 796 were homosexual or bisexual, and just under half of these had antibodies to HIV. In all, almost nine out of every ten men infected with HIV had some signs of enlarged lymph nodes. But the severity of the lymphadenopathy did not correlate with the degree of damage to their immune systems, or with the severity of their other symptoms.

Overall, it seems that the risk of someone with persistently swollen lymph glands developing AIDS during a five-year period is between 10 and 30 per cent – about the same as for all HIV-infected people. Some people with persistently swollen lymph glands are otherwise healthy. Others experience a range of symptoms of varying severity. People infected with the virus but without swollen lymph glands may also develop these symptoms. These conditions include (apart from swollen lymph glands) fatigue, diarrhoea, night sweats, weight loss of over 10 per cent, and persistent fever. These symptoms may persist or may appear intermittently, lasting for several weeks at a time.

Before the advent of the test for antibodies to HIV, doctors used to use the term AIDS-related complex (ARC) to describe these types of symptoms. Someone had ARC if they had two or more such symptoms for three months or longer, together with two or more abnormal laboratory tests, such as low levels of T-helper cells. Another reason why the term ARC is less commonly used now is that the definition of AIDS itself has changed. The new definition, formulated by the Centers for Disease Control in Atlanta, came into effect in September 1987. It included for the first time the group of symptoms known as 'HIV wasting syndrome' or, in Africa, 'slim disease'. These symptoms include severe loss of weight and diarrhoea, which may be accompanied by fever. Patients with this syndrome, formerly classified as having ARC, are now said to have AIDS.

Other problems that people infected with HIV may experience include skin rashes and viral and fungal infections. These include tinea (ringworm), thrush (a fungal infection) affecting the mouth, herpes simplex (oral or genital) and herpes zoster (shingles). Attacks of herpes are more severe and last longer than in healthy people. Tooth decay, mouth ulcers and dental abscesses may also occur. One condition, formerly very rare, which occurs in people infected with HIV, is called oral hairy leukoplakia. This appears as a white, warty area on the side of the tongue and on the cheeks inside the mouth. It may be caused by a virus.

Doctors have tried to establish whether any of these conditions can predict which patients are more likely to develop AIDS. Some studies have suggested that the appearance of oral hairy leukoplakia and oral thrush indicates that the person is at greater risk of progressing to AIDS. An earlier indicator may be the development of shingles, however. Mads Melbye of the Institute of Cancer Research in Aarhus, Denmark, together with American researchers, studied the incidence of shingles in homosexual men registered at a private practice in central Manhattan in New York. By 1983, over half of these patients were seropositive for HIV.

Shingles results from reactivation of the herpes zoster virus (also called varicella), which causes chickenpox in childhood. The infection can remain latent in nerve cells, flaring up again when immunity is low. Between January 1980 and May 1986, 112 homosexual men registered at the Manhattan practice developed herpes zoster. Three of these men had already developed AIDS and were excluded from the study. Out of the remaining 109, twenty-four (22 per cent) had developed AIDS by early 1987, and eleven of them had died.

Melbye and his colleagues found that within two years of developing shingles, about 23 per cent of the men had developed AIDS. After three years, this figure rose to about 35 per cent and after four years, to about 46 per cent. By the end of four years, over a quarter had died from AIDS. A patient's risk of subsequently developing AIDS rose if the shingles was severe and painful and if it occurred on the face or neck. The researchers added: 'Both oral thrush and oral hairy leukoplakia were likely to be diagnosed after herpes zoster, arising at an average of 1.2 and 1.1 years later, respectively.' This

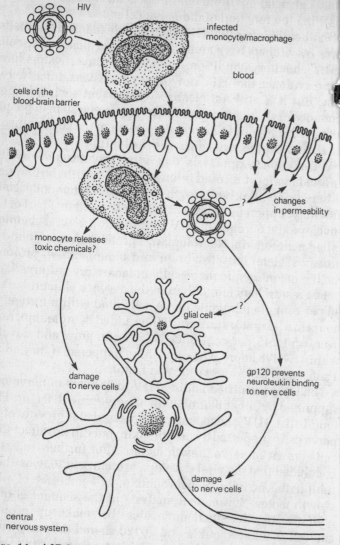

HIV

infected
monocyte/macrophage

blood

cells of the
blood-brain barrier

changes
in permeability

?

monocyte releases
toxic chemicals?

glial cell

?

damage
to nerve cells

gp120 prevents
neuroleukin binding
to nerve cells

damage
to nerve cells

central
nervous system

Figure 11. AIDS and the brain. Scientists are not sure which of these theories may account for the dementia and other neurological effects that some people with AIDS develop.

study also found that lymphadenopathy did not seem to correlate with a high risk of AIDS.

The researchers concluded:

> The incidence of AIDS after herpes zoster is approximately linear, with half of the AIDS cases arising within four years of zoster. Extrapolation of this linear trend suggests that eight years would be about the longest incubation period between zoster and AIDS. If we add to this another two to seven years between HIV seroconversion [development of antibodies to HIV] and zoster, the risk of AIDS developing after HIV seroconversion must continue for at least ten to fifteen years.

Some patients infected with HIV first visit their doctors not because they are suffering from typical infections, but because they are becoming forgetful or apathetic or are unable to concentrate or initiate and control voluntary movements. Changes in behavior may also occur. These neurological symptoms are caused by HIV attacking the brain and nervous system. About 60 per cent of HIV-infected people have symptoms of this kind, while up to 90 per cent have abnormalities of the brain and nervous system that doctors can detect at a postmortem examination. Full dementia, similar to that which occurs in Alzheimer's disease, develops in about 80 per cent of patients with neurological symptoms within a year, according to some studies. Incontinence, inability to coordinate movements and slight paralysis are common. AIDS dementia, which is also called AIDS encephalopathy, is included in the new definition of AIDS which came into effect in September 1987.

When doctors first began to recognize that many patients with AIDS also developed mental disturbances, they initially thought that these symptoms might be due to the brain infections and tumours that some people with AIDS develop, or to psychological effects. Yet research has shown that HIV itself is to blame, even though it does not commonly infect nerve cells. Many studies have now confirmed that, in many AIDS patients, it is possible to isolate HIV and viral components from the brain and the cerebrospinal fluid, which bathes the tissues of the central nervous system. In some patients, the amount of virus isolated from the brain and cerebrospinal fluid is far greater than that present in the blood.

Postmortem examinations of the brains of people who had suffered from AIDS dementia showed that the tissue of the brain had shrunk. Under the microscope, researchers saw characteristic abnormalities. In some individuals, small groups of inflammatory cells appeared throughout the brain. Other patients had no inflammation, but spaces had appeared in some parts of the brain. Researchers also often found abnormally large cells with many nuclei which they called 'multinucleated giant cells'. These cells seem to resemble the white blood cells in laboratory cultures that fuse together when infected with HIV (see p. 101). Before long, several teams of researchers had identified the infected cells in the brain as monocytes and macrophages. Other workers also suggested that the 'multinucleated giant cells' probably started out as macrophages, too, because of similarities in their structure. David Ho, of the Cedars-Sinai Medical Center in Los Angeles, and his colleagues Roger Pomerantz and Joan Kaplan, put forward a theory to explain how HIV manages to gain access to the central nervous system. First, HIV infects the monocyte in the bloodstream. The infected cell then passes across the blood–brain barrier – the tightly linked cells that keep the bloodstream separate from the fluid that bathes the central nervous system – and into the brain. These researchers then suggest two possibilities. First, the infected cells might release toxic chemicals. These substances might damage the nerve cells and the glial cells, which protect the nerve cells and manufacture the insulation around their fibres. Or, these toxins might attract inflammatory cells which carry out the damage themselves.

The second possibility is that the infected monocytes might affect the cells of the blood–brain barrier, so altering the permeability of the barrier. Such a change would alter the deli-

cately balanced environment of the central nervous system, upsetting the function of the nerve cells.

Another theory is that HIV may be able to infect glial cells. Some researchers have found that HIV can attack these cells under laboratory conditions. In addition, investigators have found evidence that HIV can, on rare occasions, infect nerve cells. But it is still not clear to what extent such infection accounts for the malfunctions of the nervous system in people with AIDS. Figure 11 summarizes some of the current theories about how the virus damages the brain.

Yet another proposal is that viral proteins released by infected monocytes could interfere directly with the function of nerve cells. Work by Ho, this time with other colleagues, provides support for this theory. They showed that part of the envelope glycoprotein of HIV is similar to a natural chemical called neuroleukin. Neuroleukin, a protein which is found in human skeletal muscles, brain and bone marrow, prolongs the life of embryonic nerve cells in laboratory cultures.

The researchers found that, in the absence of neuroleukin, 90 per cent of embryonic nerve cells died within forty-eight hours in a laboratory culture. If they added neuroleukin, however, up to 45 per cent of cells began to grow and develop within twelve hours. The same thing happened if they added a substance called nerve growth factor.

The team then tried adding HIV to cultures of nerve cells supported by either neuroleukin or nerve growth factor. They found that HIV consistently suppressed the growth of the nerve cells supported by neuroleukin, but did not affect those cells grown in nerve growth factor. After further tests, they concluded that the viral envelope protein, gp120, was able to inhibit the activity of neuroleukin, but not that of nerve growth factor. Subsequent analysis of the sequences of the amino acids, the small molecules that make up proteins, showed that neuroleukin and gp120 shared the same amino acids at fourteen out of forty-seven sites – a similarity of about 30 per cent. This similarity, the researchers said, appears in all 'isolates' or strains of HIV in which scientists have studied the sequence of the gp120 protein. In support of this work,

a short sequence of amino acids which corresponds to the sequence common to both gp120 and neuroleukin has also been shown to block the action of neuroleukin in similar experiments with nerve cells.

The scientists concluded from their results that the viral envelope protein may compete with neuroleukin for binding to nerve cells. Those cells to which gp120 binds probably die sooner, and this could account for the neurological abnormalities of AIDS.

There is one more facet to this interesting theory. Neuroleukin is also produced by T-cells that have been activated by an antigen. In this case, neuroleukin has the function of stimulating B-cells to produce antibodies. Perhaps gp120 also interferes with neuroleukin's function in activating B-cells. Such interference could, of course, account for the strange behavior of B-cells in AIDS. It might even be the primary role of this sequence of amino acids to sabotage the role of B-cells. David Ho and his colleagues conclude: 'If so, the brain may be an "innocent bystander" in AIDS, a chance target of HIV due to viral inhibition of a factor that is used by both the brain and the immune system.'

The final stage of the disease for many people infected with HIV is the syndrome of unusual infections and tumours that doctors in the US originally dubbed 'AIDS' in 1982. The two most common manifestations are pneumonia caused by the microorganism *Pneumocystis carinii* and the tumour Kaposi's sarcoma. About 80 per cent of people with AIDS in the US have either or both of these conditions.

To immunologists, the infections and tumours typical of AIDS indicate that the cells of the immune system are not working as they should. Various microorganisms take advantage of this deficiency in the individual's immune protection. The so-called 'opportunistic infections' that develop will depend largely on the types of microorganisms that a person encounters. This explains why *Pneumocystis carinii* pneumonia is so common in AIDS, particularly in the West.

People come into contact with *Pneumocystis carinii* all the time: in healthy individuals, the cells of the immune system deal with it readily, but people with severe immunodeficiency succumb to the infection.

Someone with AIDS is therefore at risk of falling ill with a whole range of infections. Viruses, parasites and fungi all take advantage of the ailing immune system. So do cancers.

The tumour that most frequently appears in AIDS is Kaposi's sarcoma. This malignancy takes the form of a purplish patch on the skin. It also occurs in the mouth and throughout the gut, although these internal tumours, like the ones on the skin, rarely cause problems in terms of the patient's health. In the US, Kaposi's sarcoma used to be rare – fewer than three cases per million men occurred every year. Yet, by the late 1980s, Kaposi's sarcoma was 2,500 times more common in young single men in the US than it was in 1980.

Other tumours common in people with AIDS include lymphoma – a type of tumour which arises in lymph nodes – and cancers of the mouth, tongue, bowel, rectum and anus. Some studies have found that men with AIDS who had anal warts also had a high risk of developing cancer of the anus.

In Africa, the range of infections and tumours is similar, although HIV wasting syndrome – known as 'slim disease' – may be relatively more common. The opportunistic infections may also appear with differing frequencies. Tuberculosis, for example, is more commonly associated with AIDS in Africa than it is in the US and Europe. *Pneumocystis* pneumonia is less common in Africa.

Doctors caring for patients with AIDS believe that the infections and tumours typical of AIDS may change as the epidemic progresses. Kaposi's sarcoma and *Pneumocystis* pneumonia may be typical of those people who fall ill relatively early in the course of infection. People who incubate the disease for longer may develop other problems: perhaps dementia, or another range of opportunistic infections. The characteristic features of the disease may well continue to change as time passes.

*

One of the factors that makes a diagnosis of HIV infection particularly difficult to cope with is that no one is sure what proportion of HIV-infected people go on to develop AIDS. According to estimates by the World Health Organization, between 10 and 30 per cent of HIV-infected people will probably develop AIDS over a period of five years, and 20 to 50 per cent will develop other symptoms and illnesses related to HIV infection. In a study of 270 men in London, about 7 per cent of those seropositive for HIV develop AIDS each year, and this rate shows no sign of levelling off. In other studies, the likelihood of developing AIDS during a three-year period has ranged from 15 per cent to over 30 per cent. The reasons for the differences in these figures may be that some groups had been infected with HIV for longer. Alternatively, the way in which the virus enters the body may be important in determining the incubation period.

One reason why it is difficult to estimate the incubation period, of course, is that many people do not know when they were first exposed to the virus. This is less of a problem for those infected via contaminated blood transfusions. By late 1987, there were over 1,170 cases of HIV infection from this cause in the US.

Two scientists, Graham Medley and Roy Anderson from Imperial College in London, together with researchers from the University of Georgia in the US, studied data from the US on people who had received infected blood. They found that the length of time that it took these people to develop AIDS varied according to their age and sex. On average, it took females 8.77 years to develop AIDS from the time of infection, whereas it took males only 5.62 years. Patients older than fifty-nine and younger than five developed AIDS more quickly. Those in the older age group took, on average, 5.5 years to develop AIDS. Young children, on average, developed AIDS in just under 2 years.

It is difficult to draw firm conclusions from such a study, however. These researchers warn that no one knows how many people received infected blood transfusions in the days before health authorities screened donated blood for antibodies to HIV. So we cannot know what proportion of infected individuals will eventually develop AIDS. In addition, it may not be possible to extrapolate these results to people who became infected via sexual activity, for example. Medley and his co-workers emphasize: 'It is also unknown how these results from an industrialized nation pertain to the developing world, where individuals are typically exposed more frequently and to a larger range of infectious agents.'

Doctors know with rather more certainty how long someone is likely to live once they have AIDS. About one in five is still alive after three years. The length of survival seems to depend on which particular condition the individual has developed. People with *Pneumocystis carinii* pneumonia survive, on average, between nine months and a year. Those with Kaposi's sarcoma have a slightly better outlook, living for two to three years after diagnosis.

One encouraging breakthrough has been the development of a test which can predict which people infected with HIV are more likely to develop AIDS. The test detects viral antigen, the core protein called p24. This protein appears in the blood for a brief period at around the time that some infected people experience a short illness resembling glandular fever or influenza, before antibodies appear. Then, as the person develops antibodies to p24, the antigen disappears (see figure 12). Later on in the course of the disease, the situation reverses again. People start to lose the antibody to p24. As levels of this antibody drop, p24 antigen starts to reappear. Research suggests that people who keep producing antibodies to p24 are more likely to remain well than those who lose these antibodies and in whom antigen reappears.

Doctors in the US have already used a test for viral antigen to diagnose early HIV infection in four previously uninfected homosexual men who had symptoms such as fever, rash, sore throat and aching joints and muscles. Normally, it is difficult to identify this initial illness as HIV infection because other viruses can cause similar symptoms and antibodies to HIV have not yet developed. In these four patients, tests for antigen were positive. All of them subsequently developed anti-

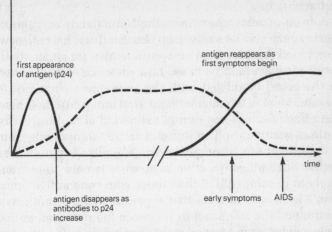

first appearance
of antigen (p24)

antigen reappears as
first symptoms begin

time

antigen disappears as
antibodies to p24
increase

early symptoms   AIDS

Figure 12. *The viral protein p24 appears in the blood soon after exposure to the virus. Then, as antibodies to p24 develop, p24 itself disappears. Some research suggests that when the p24 protein reappears in the blood, at around the time that antibodies to p24 disappear, this signifies increasing viral replication, and a greater risk of developing AIDS.*

bodies to HIV. So the test for antigen can allow doctors to diagnose infection with HIV at an earlier stage.

In addition, several studies have shown that loss of p24 antibodies and reappearance of p24 antigen is associated with a greater risk of progressing to AIDS. For example, a team of researchers from Copenhagen in Denmark, led by Court Pedersen, tested a series of 276 blood samples, taken from thirty-four men who had antibodies to HIV, for presence of the core protein of the virus, the p24 antigen. From the results, the team estimated that about half of all infected patients will develop AIDS within two years of the reappearance of viral antigen. These researchers also suggested why p24 antibodies should disappear as p24 antigen rises. They proposed that, in the later stages of infection, as replication of the virus increases, the amount of viral protein in the blood goes up.

Antibodies to p24 mop up this protein, forming antigen–antibody complexes; this would account for the loss of antibodies in the blood.

A team of researchers from the Netherlands and the US carried out a similar study. Jaap Goudsmit and his colleagues took blood samples from seropositive men for an average of about nineteen months to see how presence of viral antigen in the blood correlated with the subsequent course of the disease. Most of the men without viral antigen in their blood were free of symptoms during the period of the study. Both antigen-positive and antigen-negative individuals were equally likely to have persistent generalized lymphadenopathy. But antigen-positive men were twenty times more likely to develop AIDS than those who were antigen-negative. The length of time that a person had had antigen in their blood did not seem to influence the outcome: antigen appeared between five and eighty-two weeks before the diagnosis of AIDS.

At the time of writing, the antigen test is available only to research scientists, but it will probably become more widely used in medical practice in the future. Doctors will be able to reassure patients who are doing well, and keep a close check on those at risk of developing severe infections. The test will also play a key role in the development of new drugs. If p24 antigen disappears in all the patients given a new drug, this would suggest that the drug is preventing viral replication. Such an objective means of assessing the action of a new drug is far more accurate than asking patients how they feel and counting the number of infections they suffer. The antigen test will greatly help doctors to establish whether a drug is effective or not, speeding up the process of evaluating new therapies.

CHAPTER 9
PATHS OF SPREAD

The spread of AIDS is not inevitable. We have enough information now to prevent the virus spreading without check. This strategy relies on changing people's behaviour, which is difficult to accomplish. But it may take many years before scientists succeed in their struggle to develop a vaccine. Until then, the only way to control the epidemic will be to prevent the virus from spreading along its various routes of transmission – sexual intercourse, contact with blood and from mother to child. It is interesting to contemplate how, after observing the first cases of AIDS in the US, doctors worked out the ways in which the virus passes from one person to another. In the US, AIDS first appeared in highly specific groups of people. If coughs, sneezes, handshakes, kissing and other types of casual contact had spread the virus, a completely different pattern of infection would have emerged, with cases appearing at random throughout the population.

Studies of who was affected allowed scientists to draw conclusions about what activities *can* spread the virus. Yet it is not such a simple matter for a scientist to say conclusively that a particular activity will *not* transmit HIV. Doctors and scientists, educated in the discipline of statistics, and aware that next week someone else might conceivably prove them wrong, err on the side of caution. They will only go so far as to say that there is 'no sound evidence' for transmission by a certain route, that transmission is 'exceedingly unlikely' or that there is 'an extremely low risk' of the virus being passed on. Some people feel concerned by such cautious advice. In the real world, however, it is important to see such minute risks for what they are: unlikely. Most of the time, the risk of being knocked down by a vehicle is not enough to deter someone from crossing the road, and most people, if they take reasonable care, avoid such accidents.

There are parallels with the risks of becoming infected with HIV. Most of the time there is no risk, but there are occasions when precautions are needed in order to avoid possible infection. Some of these measures, such as screening of donated blood, are carried out by public-health authorities. Others, such as the wearing of a condom during sexual intercourse, are the responsibility of the individual.

There are times when people's fear of infection leads them to take protective measures where none is necessary. Such overreaction often stems from ignorance. Even people well versed in the evidence for and against a particular means of transmission can find it difficult to overcome their fear of such an unfamiliar disease. However, the basic research and observations on which scientists have based their advice on how HIV is and is not transmitted are revealing.

It is possible to isolate HIV from blood, semen, vaginal secretions, saliva, tears, breast milk and amniotic fluid (the fluid that surrounds the baby in the womb). Just because it is possible to isolate the virus in the laboratory, however, does not mean that the fluid concerned is capable of transmitting the infection. There are several reasons why not. The concentration of virus may be too low; the body surface that the fluid comes into contact with may not include cells susceptible to infection; and there may be natural defences which stop the virus from attacking the new host.

The main ways in which the virus that causes AIDS can be passed on are sexual intercourse and contact with infected blood. An infected mother can also pass the virus on to her baby.

During sexual intercourse, HIV can pass from man to man,

from man to woman and from woman to man. There have also been one or two reports of transmission between two females as a result of sexual contact. Although AIDS first appeared among homosexual men in the US, all the evidence suggests that any type of penetrative sexual intercourse – anal or vaginal – can transmit the virus. In parts of Africa, the main means of transmission is via conventional heterosexual intercourse. In the US and Europe, heterosexual transmission is much less common. Relatively few heterosexuals are infected by the virus, so this means of transmission would necessarily be less likely.

One factor affecting an individual's chances of becoming infected is the rate of transmission. In other words, if an uninfected person has sex with an infected person, what are the chances of becoming infected? Some researchers have studied couples where one partner is infected and the other is not. Even in long-standing relationships, despite intercourse without a condom, the uninfected partner does not always become infected. In this sort of situation, the rate of transmission of the virus seems to be highly variable. The virus seems to be passed on in between 20 per cent and 73 per cent of such couples.

No one knows, however, what the risk is if an uninfected person has sexual intercourse once with an infected partner. Case histories reported in medical journals certainly suggest that a single act of sexual intercourse with an infected person can transmit the virus. For example, women have developed antibodies to HIV following artifical insemination with semen from an infected donor. Infected women can also pass the virus on to their male partners after minimal contact: according to a case reported in 1987 in West Germany, a man became infected with HIV after having sex only twice with an infected woman.

The infection probably passes from a woman to a man when the virus, which can appear in vaginal secretions, enters the man's bloodstream via tiny abrasions on his penis. Transmission from a man to another man or to a woman is easier to explain, for normal semen contains many T-helper lymphocytes, the type of white blood cell that HIV infects. One infected T-cell can harbour thousands of viruses. If the man has any other sexually transmitted diseases, he is likely to have even greater numbers of T-cells in his semen. Interestingly, research has shown that semen can suppress the function of the cells and molecules of the immune system in the recipient. This observation ties in well with the fact that homosexual men who practice receptive, as opposed to insertive, anal intercourse are at greatest risk of HIV infection. In addition, sperm can act as antigens in the recipient, thus activating lymphocytes and macrophages. If the recipient is already infected with HIV, activation of these cells could set off viral replication. Together, these findings explain why some doctors recommend that people who are already infected should avoid further exposure to semen.

Avoiding sexual contact is of course important for people who think that they may be at risk of catching the virus or of passing it on to their partners. The use of a condom during sexual intercourse helps to avoid the transfer of body fluids.

For homosexuals, it is clear that the most risky behavior of all is unprotected receptive anal intercourse. A large study in several cities in the US, the Multicenter AIDS Cohort Study, monitored about 2,500 homosexual men for six months to see whether it was possible to identify which types of behavior were associated with the development of infection. Within this group, there were 220 individuals who had practised neither receptive nor insertive anal intercourse during the period of the study. None of these men developed antibodies to the virus. In contrast, over 10 per cent of the 548 men who had had receptive anal intercourse during the period of the study did develop antibodies to HIV. Another group of 147 men reported having receptive oral intercourse with at least one partner, but not receptive or insertive anal intercourse. None of this group developed antibodies either. This observation, together with several similar studies, seems to suggest that there is a lower risk of transmission of HIV via oral–genital contact.

The study concluded that anal intercourse was the principal route of infection. The researchers said: 'A prudent course would be to stop anal intercourse entirely.' Other researchers, more pragmatically, have recommended that if it is not possible to avoid anal intercourse, the next best thing is to use a condom, together with a water-based lubricant, in order to prevent the transfer of body fluids. A condom will not eliminate the risk of transmission altogether, but it will greatly reduce it.

As far as heterosexuals are concerned, the virus is transmitted by conventional vaginal intercourse. Anal intercourse between a man and a woman may, of course, also transmit HIV, but there is no evidence to suggest that anal intercourse is necessary for heterosexual transmission. Again, a condom will make sex safer if there is any risk that one or other partner is infected.

The second means of transmission is via blood. Doctors first identified this route in the West because some haemophiliacs began to develop AIDS. As discussed in Chapter 6, haemophiliacs need to take factor VIII to help their blood to clot. Many thousands of blood donors contribute to each batch of factor VIII, so haemophiliacs became vulnerable to HIV infection very early on in the epidemic. The development of a method of heat-treating factor VIII so that any viruses present are killed has greatly reduced the risks to recently diagnosed haemophiliacs in most countries in the West.

People who have received contaminated blood in transfusions have also become infected. This mode of transmission is still a problem in many parts of Africa, where facilities do not exist for screening donated blood. In the West, this method of spread is now extremely rare. American authorities began to screen donated blood in 1985. In Britain, all blood for transfusion is tested for antibodies for HIV. In addition, the blood transfusion service asks people in high-risk groups not to give blood. The only risk from a blood transfusion in countries where blood is screened is that someone might have given blood after becoming infected with the virus but before developing antibodies. The chances of receiving such contaminated blood are minute, however. In Britain in 1987, the blood transfusion service put this risk at one in five million. To put this figure in context, about half of the people who receive blood transfusions die within two years as a result of the injuries or illness that made a transfusion necessary in the first place.

In countries where the supply of sterile needles is plentiful, and needles are used once and then thrown away, there is no risk of catching HIV from injections or from giving blood. In some parts of the world, however, needles are scarce and may be reused for several people. Even though this practice ought to be avoided, many countries do not have the resources to pay for disposable needles and syringes. Some researchers have suggested that this route of transmission might partly account for the spread of the infection in Africa. Injections are commonplace in many parts of Africa, whether prescribed by doctors or by traditional medical attendants. Some ethnic groups also still practise scarification rituals. Conceivably, the instruments used to produce the scars on the skin could spread infection. Researchers have yet to establish precisely the role of exposure to unsterile needles or other instruments.

While it might be theoretically possible to become infected with HIV following ear-piercing, acupuncture, electrolysis and tattooing, no cases have been reported of infection by these routes. Of far greater concern in the West is the spread of HIV infection among people who inject drugs such as heroin. In some areas, for example Edinburgh in Scotland, over half of the drug users have antibodies to HIV. Drug users are at risk of becoming infected with HIV if they share a needle with someone who already has the virus. People can avoid spreading the virus by this route by using a clean needle and syringe, and not sharing needles with others.

As well as blood, donated organs and skin grafts can carry HIV. There have been a few cases documented of people who became infected in this way, but infection by this route is very rare. Except in extreme emergencies, doctors now test organ donors for antibodies to HIV before carrying out a transplant or skin graft.

The third main method of transmission is from mother to

child. An infected woman can pass the virus on to her child during pregnancy, at birth or with her breast milk during breast-feeding. In the West, information on the risks of passing on the virus during pregnancy or at birth is hard to come by because so few heterosexuals are infected. One study, in Edinburgh, of children born to women who are intravenous drug users, found that the mother passed on the infection to her child in about 50 per cent of cases. Scientists in Africa, however, have suggested that the risk may be lower, around 25 per cent.

Researchers from Project SIDA (SIDA is French for AIDS) in Kinshasa, Zaîre, and colleagues from the US, studied women attending hospital in Kinshasa. Almost 6 per cent of 6,000 pregnant women tested for HIV antibodies were seropositive. About half of the seropositive women had AIDS. The researchers tested seventy children born to infected mothers for a type of antibody which would suggest that the children were also infected. (Some types of antibodies can cross the placenta but their presence does not mean that the child is infected with the virus.) Only seventeen (24 per cent) of these seventy babies were positive in this test.

Seven months later, 18 per cent of the babies born to infected mothers had died, compared with only 1 per cent of children born to seronegative mothers. The same study found that a woman was more likely to pass on the infection to her child if she had AIDS or severe disease related to HIV infection than if she was infected but had no symptoms.

This study does not take into account the potential risk of transmission from breast milk if the child escapes infection while in the womb or during birth. It is difficult to separate these risks from each other, but there have been a few reported cases where breast milk was the only possible source of infection. For example, researchers led by Philippe Lepage, working in Kigali in Rwanda, reported the cases of two children who may have become infected by breast milk. The first case was in a child born to healthy Rwandese parents. The mother lost a great deal of blood during the birth and had a blood transfusion – her first ever – a day later. The child fell ill at the age of ten months and died nine months later. She had antibodies to HIV. The mother had been breast-feeding her daughter. In the past, people have suggested that transmission of the virus could occur if the mother has cracked nipples. But in this case there had been no nipple problems.

The mother began to fall ill about a year after the birth. She had antibodies to HIV, although her husband did not, and she denied having had sexual contact with anyone other than her husband. When the researchers traced the two people who had provided the blood for her blood transfusion, one of them was positive. The mother must have passed the virus on in her breast milk some time after the birth.

The second case reported by this group of researchers was almost identical. Such reports have led some researchers to suggest that banks of human breast milk should be pasteurized. This process inactivates the virus. But the doctors in Rwanda emphasized that their report should not discourage mothers from breast-feeding, even in countries where HIV infection is common, as breast milk is best for the baby.

As scientists have built up a picture of how HIV passes from person to person, they have also managed to identify situations in which transmission does not occur. There is no evidence to suggest, for example, that contact with saliva or tears from an infected person poses any risk. It is true that it is possible, in the laboratory, to isolate the virus from saliva and tears, but one study puts this observation into context. David Ho, with colleagues from Massachusetts General Hospital in Boston, tried to isolate the virus from 83 samples of saliva taken from 71 homosexual men who had antibodies to HIV. Fifty of the men also gave samples of blood at the same time.

The researchers were able to isolate the virus from 28 of the 50 blood samples, but from only one of the samples of saliva. This sample came from a man with pneumonia and other signs of severe disease. The culture of saliva began to show detectable signs of viral activity (the presence of the viral enzyme reverse transcriptase) only after twenty-one days, yet

126

this man's blood sample showed evidence of viral activity on day three. Ho and his colleagues concluded that the virus is present only infrequently in saliva; when it is present, it occurs in very small amounts. In addition, one study has suggested that there is some factor in saliva that prevents HIV from infecting cells.

The issue for most people is whether they can become infected by, say, drinking from the same glass as an infected person, by sharing facilities such as a toilet, or perhaps by shaking hands. The evidence that ordinary social interaction of this kind presents no risk to uninfected people comes from several studies of the household contacts of infected people. Doctors in the US, led by Gerald Friedland from New York, tested 101 people who had lived in the same households as 39 adults with AIDS. None of these 101 were sexual partners of the patients with AIDS, but all had lived with an infected person for at least three months, and, on average, for about twenty-two months. Most of the families were poor and lived in crowded conditions, sharing items such as toothbrushes, towels, plates and drinking vessels, as well as facilities such as beds, toilets and baths or showers. The study showed that only one of the 101 household contacts was infected with HIV – a child who had probably been infected from her mother at around the time of birth.

Figures from the Centers for Disease Control support the results of Friedland and his co-workers. According to the CDC, except for sexual partners, no one in the families of more than 12,000 people with AIDS is known to have developed the disease. The CDC cited five other studies which have also failed to find any transmission of HIV to adults who were not sexual partners of patients with HIV infection, or to children who were not at risk of transmission at the time of birth.

Another reassuring study was carried out in France, at a boarding-school for over a hundred haemophiliacs, epileptics and diabetics. All the children shared dormitories, classrooms, swimming-pools, dining-hall and lavatories. At the end of three years, half of the haemophiliac children had

127

developed antibodies to HIV as a result of receiving contaminated factor VIII to help their blood to clot. But none of the non-haemophiliac children had antibodies to HIV.

Some people have been concerned about whether it would be possible to catch HIV infection from public swimming-pools. In a properly maintained and disinfected pool, this is extremely unlikely. Any virus that did enter the water, from a cut on an infected person, say, would immediately be greatly diluted. In addition, chlorine kills the virus. Furthermore, if the pool were not properly disinfected, the risks of catching hepatitis B, meningitis or polio would be far higher than that of picking up HIV. There have been no cases to suggest that it is possible to become infected from a swimming-pool.

One unfounded theory which has caught the imagination of many people is the question of whether blood-sucking insects could transmit HIV from person to person, in the same fashion as mosquitoes spread malaria. There is a great deal of evidence against this idea. One of the most convincing observations is the pattern of spread of the virus. In Africa, malaria is more common in children than in adults, which suggests that mosquitoes bite children more often than adults. Yet children rarely have HIV infection, unless they were born to an infected mother, or are old enough to have caught the infection by sexual contact. HIV infection is also much less common in rural areas than in towns in Africa. If mosquitoes spread HIV, the opposite would be true. And in Africa, as in the US and Europe, studies have shown that people sharing the same household as an infected person have no increased risk of the disease, unless they are a sexual partner or a young child of the infected individual. As many researchers have said, it is an odd mosquito that prefers to feed on prostitutes and dislikes feeding on children.

Why should mosquitoes be able to transmit malaria but not HIV? When a mosquito takes a meal of blood from an infected person, the parasite that causes malaria enters the mosquito's salivary glands, where it multiplies. When the mosquito feeds subsequently, it injects an anticoagulant – along with more malarial parasites – to keep the blood of its victim flowing.

Unlike the organism that causes malaria, HIV is unable to replicate in the cells of insects.

If HIV cannot survive in the mosquito, the only other way in which such blood-sucking insects could theoretically transmit the virus is mechanically: the insect's mouthparts would have to act as a tiny hypodermic syringe. Researchers have explained why transmission in this way is very unlikely. Although insects can transmit some viral diseases, in these cases the concentration of virus or of infected cells in the blood is very high. By contrast, only one in a million lymphocytes is likely to be infected with HIV – and, as we have seen, many people with AIDS have very low numbers of lymphocytes anyway.

Thomas Monath, a virologist with the Centers for Disease Control, has studied in some detail the issue of whether mosquitoes and other blood-sucking insects can transmit HIV. The CDC sent researchers to a town called Belle Glade in Florida, after allegations that the town had an exceptionally high incidence of AIDS, comparable to that in the areas of highest prevalence in the United States, San Francisco and New York City. Some scientists suspected that mosquitoes were to blame. But the researchers found that high-risk behaviours such as use of intravenous drugs, prostitution and multiple sexual partners were common in the town. In addition, most cases were in young adults; there were no cases in children or elderly people. So there is no evidence here that mosquitoes can transmit HIV.

Monath has managed to 'infect' another blood-sucking insect, the bedbug, with HIV. The virus can survive in blood that remains on the insect's mouthparts or in its gut. But the chances of a bedbug transmitting the infection are minute, for the following reasons. Bedbugs can transmit two other diseases, called equine infectious anaemia and bovine leukosis. Both of these infections are caused by retroviruses, the family of viruses to which HIV belongs. Yet in both cases, the amount of virus that appears in the blood of the infected horse or cow is very high. Monath says that for a bedbug to transmit these diseases, the blood must contain over one million viral particles per millilitre. The blood of people with AIDS, however, usually contains less than ten viral particles per millilitre.

Another factor to consider is how much blood the insect can hold on its mouthparts. The volume of blood that a bedbug can transfer is of the order of one fifteen-thousandth of a millilitre. Monath calculates that the amount of blood on a hypodermic needle is about 140 times the amount of blood on the mouthparts of a bedbug. Yet, of the many health-care workers who have accidentally stabbed themselves with a contaminated needle, only a handful have become infected.

Health-care workers do, of course, have to take special care when handling blood which may be infectious. By the end of April 1987, the Centers for Disease Control had received reports of 332 health-care workers who had been exposed to the blood or other body fluids of people infected with HIV. These cases included 103 needlestick injuries (where people had stabbed themselves with a needle) and 229 instances where blood had splashed into the worker's mouth or eyes. None of these people developed antibodies to the virus. Two other studies, one British and one American, involving a total of 279 people with similar exposure to the blood of infected patients, also failed to identify any cases where the virus was transmitted.

These three studies show that the risk to health-care workers is low. Out of the thousands of cases of accidental exposure, by late 1987, there were only a dozen or so cases of people who had developed HIV infection as a result. Four of these people had suffered needlestick accidents; two more had had extensive exposure to the blood or other body fluids of infected patients. Neither of these individuals had observed recommendations for methods of reducing the risk of infection when caring for patients.

The CDC released the details of the next three cases just before the Third International Conference on AIDS in Washington DC in June 1987. Many delegates were shocked and surprised by these cases, for none of the three had suffered a breach of the skin such as a needlestick injury or a cut. The first case was a health-care worker who had applied pressure to a bleeding site on a patient in an emergency room. Her hands were chapped and may have been in contact with contaminated blood for as long as 20 minutes. She wore no gloves. In the second case, contaminated blood had splashed a health-care worker on the face and in the mouth. She was wearing gloves and protective glasses. She had acne but no open wounds, and washed the blood off immediately. She had also subsequently scratched herself with a needle contaminated with blood from an intravenous drug user who may or may not have been infected with HIV.

Thirdly, a technician became infected after blood from an infected patient covered her arms and forearms during an accident with some equipment. She washed within minutes. This woman said that she did not remember having any open wounds on her hands, but she had dermatitis on one ear, which she may have touched.

The CDC's publication, *Morbidity and Mortality Weekly Report*, says that although the exact route of transmission is not known, these three cases suggest that exposure of skin or mucous membrane may very occasionally result in transmission of the virus. The risk, the report adds, is likely to be far lower than that associated with needlestick injuries.

A British study published in 1987 put the risk of developing HIV infection following a single accidental exposure at 'probably less than 1 per cent'. Marian McEvoy and colleagues in London said that there had been only seven cases of occupationally acquired infection in which the people were shown to seroconvert as a result of the exposure. In a further five cases, health-care workers were found to be seropositive after exposure to blood of infected patients, but there was no evidence about whether they had been seronegative before the exposure.

Since these cases, there were two further reports in late 1987 of infection with HIV as a result of occupational exposure. One was a laboratory worker who became infected with a strain of the virus indistinguishable from the one with which the person was working. The second was an employee at San Francisco Hospital who suffered a stab with a needle containing blood from a patient with AIDS.

Some people who work in the health services, including doctors, have called for all people seeking medical care to be screened for antibodies to HIV. Yet many medical authorities believe that such demands are misguided. Apart from the problems of needing to obtain consent for such tests, screening of this kind would not be an effective way of protecting health-care workers. In many situations, especially emergencies, the result of the test would not be available rapidly enough for staff to act on the information supplied. And a negative result might provide a false sense of security: the person might be infected but not yet have developed antibodies. The practical and ethical considerations involved in caring for patients who may be infected with HIV are among the many problems that society has yet to solve in relation to the spread of the human immunodeficiency virus.

CHAPTER 10

QUEST FOR A CURE

Infection with the human immunodeficiency virus poses an exceptionally difficult problem for scientists who hope to cure the disease. The virus integrates its own genetic material into the cells of its host. To destroy those foreign genes would mean eradicating all infected cells. Earlier on in the epidemic, this option might have seemed feasible. HIV infects the blood cells known as T-helper lymphocytes. These cells have a limited lifespan, and the bone marrow continuously produces new blood cells to replace the old ones. Perhaps, scientists wondered, it might be possible to wipe out a whole generation of T-helper cells, eliminating the virus in the process.

Then it became clear that HIV also attacks other cells, including cells in the brain. The prospects of ever ousting the virus from the body receded further. Killing brain cells in order to eradicate HIV would be a rather drastic treatment. In addition, even if scientists could control the damage that the virus causes to the immune system the brain could still act as a reservoir of infection from which fresh virus could spring. It is also now certain that many of the neurological symptoms of AIDS and HIV infection are the result of the effects of the virus on the brain and nervous system. In order to have any influence on this aspect of the disease, antiviral drugs would have to penetrate the blood–brain barrier, the tightly linked cells that line the blood vessels in the brain, and preserve the delicately controlled environment which nerve cells in the brain need.

Although there is little prospect of ever being able to eliminate the infection from the body, this does not exclude the possibility of developing a treatment. The history of medicine suggests, however, that progress in treating AIDS is more likely to come about by a series of increments rather than in a single breakthrough. The discoveries of penicillin and subsequently of other antibiotics are notable exceptions.

The quest to find a cure for cancer is a good analogy to the pursuit of drugs to conquer AIDS. Although doctors cannot talk about curing cancer, they have found out how to extend the life expectancy of people who develop some types of cancer. Many women treated for breast cancer, for example, survive the disease for many years. Studies have shown that, provided the woman seeks treatment early, the eventual cause of death is often unrelated to the cancer. Medicine has gained such advances not in giant leaps but by gradually chipping away at the problem, little by little.

The treatment of AIDS may benefit from a similar approach. Although much research is currently focused on antiviral drugs, appropriate treatment of opportunistic infections may be equally important. Whether medical scientists are neglecting this field of research is a matter for debate. Certainly, in 1987, at the Third International Conference on AIDS held in Washington DC, research into ways of combating the infections and tumours that characterize AIDS was notable by its absence. Some of the finest minds in science seemed to be concentrating on the details of how the virus replicates with a view to a long-term solution. While such basic science is essential, medicine must continue to press ahead with the painstaking methods by which so many advances have been made in the past.

One example of how suitable treatments can improve the quality of life for people with AIDS has come from doctors in San Francisco. Many people with AIDS suffer repeated attacks of pneumonia caused by the microorganism *Pneumocystis carinii*. Two antibiotics are effective against this infection, but both have side effects that demand a change of drug in up to half of all patients. The doctors decided to try a method that would deliver the drug direct to the lungs, using a device which turns the drug into a very fine spray which the patient inhales. The drug, pentamidine, stayed where it was needed – in the lungs. As a result, unpleasant side effects on other organs were avoided. Other researchers have reported that preventive treatment with antibiotics can cut the number of attacks of this pneumonia. Advances such as these may seem small in scientific terms, but they may greatly improve the quality of life for the sufferers.

Of course, many of the infections that occur in AIDS are caused by other viruses, including herpes viruses and cytomegalovirus. Treatments for these infections are either non-existent or in their infancy. Despite years of effort, the lack of antiviral drugs against these infections testifies to the difficulty of the task that lies ahead in developing drugs against HIV.

Given the short history of the disease, the development of at least one drug that prolongs the lives of people with AIDS is an immense achievement. Zidovudine (formerly known as AZT), though it has some drawbacks and is exceptionally expensive, has given both public and scientists grounds to believe that drugs may work in HIV infection. Zidovudine has borne out the hopes of people with HIV infection: despite its side effects in some patients, there is little doubt that it can prolong life.

The efficacy of other drugs is less clear. The attempts of researchers to evaluate promising new preparations have been hampered in some cases by patients understandably desperate to try new drugs. Particularly in the US, there is a black market in drugs for the treatment of AIDS, even though these drugs are of uncertain usefulness. Many people with AIDS have turned to 'kitchen sink' recipes of therapies in their desperation to try some form of treatment. Doctors and scientists have not always helped. Researchers have sometimes announced their results prematurely, without scientific data to back up their claims. For example, three French

135

researchers – Jean-Marie Andrieu, Philippe Even and Alain Venet – at the Laennec Hospital in Paris held a press conference in 1985 to tell everyone that they had seen spectacular improvements in the condition of two patients with AIDS. These patients had been taking the drug cyclosporin for about a week – hardly long enough to make a balanced assessment of the treatment. Sadly, subsequent investigations showed that their optimism was unfounded.

On other occasions, companies and researchers seem to have paid more attention to ensuring that the stockmarket knew about their results than to informing the medical community. The motive, no doubt, was to boost their share prices.

The lack of novel drugs to treat AIDS and HIV infection has certainly not been for want of trying. Each new drug to undergo tests has raised hopes – only, so often, to dash them again. In July 1985, the actor Rock Hudson flew to Paris for treatment with the drug HPA-23. At the time, Hudson was probably the most prominent American to acknowledge openly that he had AIDS. In the event, the drug was of little help. Hudson died on 2 October 1985.

HPA-23 is one of a group of drugs that acts on the enzyme reverse transcriptase. This enzyme is responsible for copying the viral RNA into DNA. The DNA then inserts itself into the DNA of the host cell. In 1985, Willy Rozenbaum of the Pitié Salpétrère hospital in Paris reported the results of giving HPA-23, which contains the chemical elements antimony and tungsten, to four patients. In one, the virus stopped multiplying and the patient suffered no opportunistic infections for a year after the end of therapy. Improvements in the condition of the other three were less clear cut, even though the drug appeared to halt the replication of the virus. The side effects of HPA-23 have turned out to be too severe, however, for doctors to persevere with this treatment.

A second drug, called suramin, showed similar early promise. Suramin also inhibits reverse transcriptase. This drug was originally developed in the 1920s to treat sleeping sickness (trypanosomiasis) in Africa. In 1985, Samuel Broder, of the National Cancer Institute in Bethesda, Maryland, reported some of the first results of using suramin to treat AIDS. His team treated ten patients for six weeks. The researchers had managed to isolate virus from four of these patients before the trial and, in these four, levels of the virus fell during treatment. American scientists set up a large study involving several hospitals to determine whether suramin was indeed effective.

They published their results in September 1987 in the *Journal of the American Medical Association*. They found that, with higher doses of suramin, greater suppression of the virus seemed to occur. Unfortunately, this suppression did not correlate with improvements in the patients' immune systems or general condition. In fact, none of the patients improved immunologically, and many of them deteriorated in this respect. Although suramin caused toxic side effects, these were reversible and rarely severe.

The researchers suggest several reasons why the drug did not improve the condition of the immune systems of these patients. First, suramin merely slows or inhibits the growth of the virus: it is a 'virustatic' drug. Possibly, they speculate, some lymphocytes continued to become infected. If this is true, it may mean that virustatic drugs are of little use in the later stages of the disease. Secondly, the drug does not penetrate well into the central nervous system. This failure, say the researchers, 'allows for a sanctuary for continued viral replication and eventual re-seeding of the circulation'. Their paper concludes: 'Suramin . . . is not recommended as therapy for AIDS.'

A third drug, phosphonoformate (trade name Foscarnet), has had encouraging results from initial tests on patients. Doctors in London decided to conduct a small trial after a single patient got better when treated with the drug. This person, who had antibodies to HIV, was suffering from encephalitis (inflammation of the brain).

Phosphonoformate, like HPA-23 and suramin, can inhibit reverse transcriptase in the laboratory. To find out whether

137

the drug also works in humans infected with HIV, the team of doctors treated eleven patients with AIDS or other severe symptoms due to HIV infection with a three-week intravenous infusion of phosphonoformate. The researchers tried to isolate virus from these patients before the trial began and at regular intervals afterwards. They managed to detect virus before treatment in all eleven patients. After therapy, they were unable to isolate virus from six of the eleven. Overall, they managed to isolate virus in one in five attempts from the patients treated with phosphonoformate. In comparison, in four other patients with similar symptoms who received no phosphonoformate, they found virus on seven out of ten occasions.

The treated patients did suffer some side effects, but these were not severe. The team concluded: 'The results of this study suggest that a long-term trial of phosphonoformate is indicated in HIV-infected symptomatic patients, particularly if a successful oral preparation becomes available.'

Unfortunately, there was no improvement in the immune systems of the patients who took part. This, said the researchers, is 'disappointing but not surprising'. They added: 'It is perhaps reasonable to assume that many weeks or months of virustasis [inhibition of the growth of the virus] may be needed to achieve detectable improvement in the immune system . . . If reverse transcriptase inhibitors are to prove a successful treatment for HIV disease, therapy may need to be lifelong and may need to be given before significant deterioration in cell-mediated immunity has occurred.' Swedish researchers who carried out a similar small trial also obtained encouraging results.

Only further studies, in the form of a larger, properly controlled trial, will show whether phosphonoformate will provide a useful treatment for AIDS. The ideal drug for this infection can be taken by mouth, is non-toxic when taken over long periods, and is able to cross the blood–brain barrier. The main drawback with phosphonoformate is that, at present, it has to be given intravenously. There is evidence, however, that phosphonoformate can penetrate into the brain. Studies have shown that levels of the drug in the cerebrospinal fluid, which bathes the central nervous system, can reach 25–80 per cent of the levels in the blood. Tests of phosphonoformate are also continuing in the US.

Researchers have yet to prove the worth of phosphonoformate, and have more or less abandoned HPA-23 and suramin. But while scientists are bound to draw some blanks in the search for effective drugs, they know that the strategy of inhibiting reverse transcriptase can be successful. Zidovudine, the drug which became famous as AZT, has proved that for them.

Zidovudine (trade-name Retrovir) started life with the cryptic label 'Compound S'. Scientists had first isolated it in 1964 during a search for drugs that would be effective against cancer. Twenty years later, the American pharmaceuticals company, Burroughs Wellcome, during a trawl for compounds that could combat the virus that causes AIDS, found that Compound S could inhibit retroviruses in the laboratory.

The compound was azidothymidine or, to give it its full chemical name, 3'-azido-3'deoxythymidine. Its correct generic name is now zidovudine – an essential change to avoid any confusion with the immunosuppressant drug azathioprine. Tablets of azathioprine carry the initials 'AZT' and it would be extremely unfortunate if people were to mistake this drug for zidovudine.

Zidovudine is a member of a class of drugs called nucleoside analogues. Nucleosides are the chemicals that make up the backbone of the DNA molecule. Each nucleoside contains a sugar as well as one of the four bases that code for the genetic information in DNA. Zidovudine is an analogue of the nucleoside thymidine (see figure 13). Both contain the base thymine, but zidovudine contains a modified sugar.

When HIV infects a human cell, the viral enzyme reverse transcriptase controls the formation of viral DNA, using viral RNA as a template. The DNA molecule grows as reverse transcriptase adds new nucleosides to it. Unlike thymidine,

139

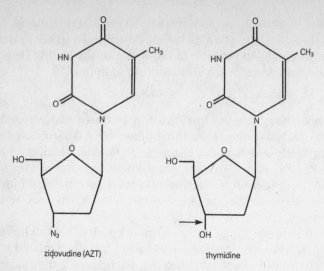

zidovudine (AZT)   thymidine

Figure 13. Zidovudine (left) is very similar in structure to the naturally occurring nucleoside thymidine (right). Where thymidine has
an -OH group, however, zidovudine has a $-N_3$ group.

however, once zidovudine has attached itself to the growing
DNA chain, the next nucleoside cannot join it (see figure
14). So zidovudine blocks the process of DNA synthesis by
terminating the chain.

All cells have an enzyme, called DNA polymerase, that
controls the normal manufacture of new molecules of DNA
by addition of nucleosides. There is some evidence that
reverse transcriptase accepts nucleoside analogues more
readily than DNA polymerase. This preference explains why
zidovudine affects the replication of HIV in infected cells to
a greater degree than it does in the replication of uninfected
cells. Zidovudine and similar drugs inhibit viral replication
at concentrations ten to twenty times lower than those at
which they would kill normal human cells.

Following the discovery that zidovudine could inhibit viral
replication in the laboratory, the next step was to test the drug
on patients. In the development of new drugs, all preparations first have to undergo so-called Phase I trials. These

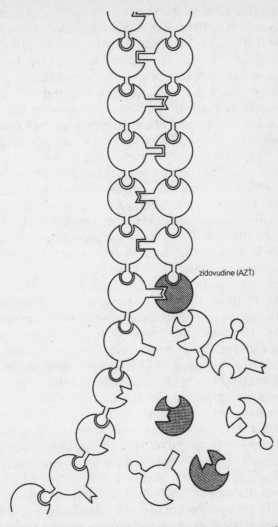

zidovudine (AZT)

Figure 14. The difference in the structure of thymidine and zidovudine means that once zidovudine has been incorporated into the DNA molecule, the next nucleoside cannot join the growing chain.

tests establish the safety of the drug, how well people tolerate it, and how quickly the body eliminates it. The Phase I trial of zidovudine showed that the drug was well absorbed when taken by mouth, and that it does cross the blood–brain barrier.

Phase II trials, to determine the effectiveness of the drug, began in February 1986. Doctors in twelve centres in the US started enrolling patients in a carefully controlled trial. The patients who took part either had AIDS, and had recently recovered from their first episode of *Pneumocystis carinii* pneumonia, or they had other severe symptoms related to HIV infection. The design of the trial allocated the participants at random to either zidovudine or a placebo preparation. In addition, neither doctors nor patients knew who was receiving the drug and who was taking the placebo, to eliminate any bias in reporting symptoms or side effects. This type of trial is called a 'randomized, double-blind, placebo-controlled trial', and it is the most powerful investigation available to doctors who want to determine the benefits and drawbacks of drugs and procedures.

The beauty of such trials is that the potential risks for those taking part balance the potential advantages. Those patients taking the active preparation may well benefit from the drug. However, they are also exposed to potentially harmful unknown side effects. Patients taking the placebo will derive no benefit from the drug – if it is indeed effective – but they will in turn be protected from unwelcome side effects.

Patients taking part in the first trial of zidovudine might therefore have been concerned, depending on their point of view, either that they were not receiving a treatment that they perceived might be useful to them, or that they might be suffering unacceptable side effects from a powerful drug. With this anxiety in mind, the National Institute of Allergy and Infectious Diseases in Bethesda, Maryland, set up a group of six people, independent of both the manufacturers of zidovudine, Burroughs Wellcome, and the researchers carrying out the trials, to monitor the results. This group, called the Data and Safety Monitoring Board, reviewed the results as they came through, first in August 1986 and again in September.

On 18 September 1986, the board recommended that, on ethical grounds, the trial should no longer be placebo-controlled. In other words, patients who had previously been taking the placebo should in future receive zidovudine. Far more people had died in the placebo group than in the group taking zidovudine.

Doctors from the twelve centres taking part in the trial published their final results in July 1987 in the *New England Journal of Medicine*. Of the 282 patients enrolled, 145 received zidovudine and 137 took placebo. After six months, only one patient had died in the group receiving zidovudine. By contrast, there had been 19 deaths in the group taking placebo. Forty-five individuals taking placebo had developed opportunistic infections, compared with 24 receiving zidovudine. In addition, the number of T-helper cells in the people in the treated group had risen.

By April 1987, after nine months of treatment, 6.2 per cent of the original group taking zidovudine had died. In comparison, out of the former placebo group, excluding those patients who were severely ill when they began treatment, and died within a few weeks, the mortality rate was 39.3 per cent. By June 1987, some patients had been taking part in the trial for over a year. Of those who had been taking zidovudine all the time, 10.5 per cent had died. Too few of the former placebo group were still alive, however, to give a meaningful mortality rate for this set.

These results were impressive. Yet zidovudine has some significant drawbacks. One is toxicity. The most serious side effect is often suppression of the bone marrow, causing anaemia. The rapidly dividing cells of the bone marrow are particularly sensitive to substances that interfere with DNA synthesis. In the trial described above, 24 per cent of patients receiving zidovudine developed anaemia, compared to 4 per cent of those taking placebo. Twenty-one per cent and 4 per cent respectively in these groups needed several transfusions of red blood cells. In addition, those on zidovudine reported feeling sick or suffering aching muscles, insomnia and severe headaches more often than patients in the placebo group.

Doctors involved with the trials warned, in the same issue of the *New England Journal of Medicine*, that '. . . the drug should be administered with caution because of its toxicity and the limited experience with it to date'. Other doctors who use zidovudine to treat patients with AIDS have reported that as many as 40 per cent are unable to tolerate it after about six weeks because of anaemia.

Some researchers have criticized the premature termination of the initial trial. They believe that the opportunity of finding out about the long-term effects of the drug, compared to a placebo, has been lost for good. Champions of zidovudine, however, say that this criticism speaks for itself. If the drug was no good, no one would be talking about its long-term side effects.

The second drawback is cost. In 1987, zidovudine was the most expensive drug on the market. A year's course for one person cost $10,000 or £7,000. Why is the drug so costly? The raw material for zidovudine is the naturally occurring nucleoside thymidine. Burroughs Wellcome used to obtain this compound from herring sperm, but the demand for zidovudine is now so great that the company buys synthetic thymidine from a contractor. It is often the rule that it is cheaper to make large quantities of a drug, but Burroughs Wellcome has found it difficult to scale up production of zidovudine. One reason is that two of the steps in its manufacture are highly explosive.

Burroughs Wellcome has said that it spent about $80m on developing the drug and expanding facilities to manufacture it. The cost, the company said in 1987, was still rising because development and testing were continuing. The search for the next generation of drugs is already under way.

Just how health services in developed countries will handle the demand for such an expensive drug remains to be seen. Some patients in the trial described had severe symptoms due to HIV infection, but had not at that time developed full-blown AIDS. The results suggested that zidovudine postponed the progression to AIDS in many of these patients. In 1987, a large trial involving 1,500 patients began in the US to determine whether zidovudine can help to delay the onset of AIDS in people infected with HIV. If it does, with an estimated 1–1½ million people infected in the US, the market for the drug could be massive. Yet, in the US, many insurance policiies do not cover the cost of drugs. People who want zidovudine but cannot afford it have to seek medical treatment at public hospitals.

In Britain, the cost of zidovudine for the treatment of AIDS has already put a strain on resources. In 1987, most of the health authorities which cared for significant numbers of AIDS patients were in London. In one health authority, the projected expenditure on zidovudine for the financial year 1988/89 stood at more than half of the authority's total budget for drugs. The government has given authorities such as this one extra funds to help them to pay for the care of AIDS patients. However, if zidovudine is proved to be useful in delaying AIDS in those who have antibodies to the virus but no symptoms, thousands of people who may have been exposed to the virus but who currently choose not to be tested will want to find out if they are indeed infected. In Britain, where in 1987 there were an estimated 50,000 to 70,000 people who were infected but did not know it, the National Health Service could have some very unpalatable decisions to make about who should receive treatment.

Zidovudine is just one of a family of drugs called nucleoside analogues. Researchers at the National Cancer Institute in Bethesda, Maryland, have already tested on humans a second nucleoside analogue called dideoxycytidine. In the laboratory, dideoxycytidine halts the replication of the AIDS virus at about one tenth of the dose required with zidovudine. Other nucleoside analogues under investigation include compounds called cyanodideoxythymidine, dideoxyadenosine and dideoxyinosine. Biochemists could produce about ten compounds, all of which might be active against the virus, by adding different chemical groups to nucleoside bases.

In May 1987, doctors at four hospitals in the US began the

145

first tests of dideoxycytidine in people. Within a couple of months, some of the people taking the drug began to complain of pains in their feet. The pains were caused by a condition called peripheral neuropathy. Fortunately, the pains disappeared when doctors withdrew the drug. Some participants also developed unpleasant skin rashes.

The investigators went on to test dideoxycytidine on another group of people, this time using lower doses. They want to find out what is the highest dose that they can use without causing side effects, and whether this dose modifies the course of the disease. Doctors also plan to test alternating therapy: one week on zidovudine followed by one week on dideoxycytidine, and so on. The theory is that such a regime might avoid the side effects that result from each of the two drugs given alone.

An interesting feature of zidovudine, dideoxycytidine and their relatives is the way in which they become active in the cell. In normal DNA synthesis, each successive nucleoside, in order to add itself to the growing chain of DNA, needs to be in a form known as a triphosphate. In other words, it has to have three chemical groups known as phosphates. Without these phosphate groups, the nucleoside cannot take part in the reaction that attaches it to the DNA. Zidovudine has to undergo the same process. Enzymes in the cell turn the drug into its active form, zidovudine triphosphate.

The need for cellular enzymes to activate these drugs is a characteristic that researchers may be able to exploit in designing new drugs of this nature. One of the problems that pharmacologists have to overcome is that of selectively targeting the drug on infected cells. This approach would avoid the problem of toxicity, where the drug unnecessarily poisons other tissues and organs. Researchers are trying to find mechanisms, specific to infected cells, which would selectively activate the drug. For example, the virus might induce novel enzymes in infected cells. However, any drug that depends on cellular, rather than viral, enzymes for its activity also has the advantage that the virus is unlikely to become resistant to the drug as a result of mutations of the virus.

146

The strategy of inhibiting the viral enzyme reverse transcriptase is an obvious one to try, for this molecule is specific to the virus. No animal cells have this enzyme. Without it, the virus cannot replicate; the genetic material of the virus would soon be broken down in the cell, its genetic information lost. Designers of drugs are therefore concentrating on different ways of inactivating reverse transcriptase.

Researchers have already determined the sequence of amino acids in the enzyme. The next step is to work out the molecule's three-dimensional structure. To do this, scientists first have to prepare the chemical in the form of a crystal. Then they use X-rays to examine the structure of the crystal. Scientists at Wellcome's laboratories in Britain announced in 1987 that they had crystallized reverse transcriptase. The next stage was to examine its three-dimensional structure. Armed with that information, the aim is to design drugs that can inactivate reverse transcriptase, perhaps by binding to the part of the enzyme crucial to its activity.

There are, however, other vulnerable points in the life cycle of the virus. It may be possible to prevent the virus from binding to the T-helper cell, for example. This could be achieved in several ways. Highly specific antibodies might block the binding site on the viral envelope protein or the receptor molecule on the cell (see also p. 181) Alternatively, scientists could design small fragments of protein called peptides to accomplish this task.

One potential therapy which relies on the principle of blocking the binding site on the cell is called peptide T. A peptide is a short chain of amino acids, the building blocks that make up proteins. Peptide T is a chain of eight amino acids, five of which begin with the letter T, hence the name. Like so many other aspects of AIDS research, the short history of peptide T has been dogged by controversy. Fuelled no doubt by elements of scientific competition, clashes of personality and political manoeuvring, as well as genuine uncertainty about the scientific basis for the claims made about this substance, the dispute about peptide T has run for many months.

Candace Pert and her colleagues at the National Institute of Mental Health in Bethesda, Maryland, described peptide T and its ability to block infection of T-helper cells by HIV in December 1986. They said that the sequence of amino acids in peptide T also occurs in the viral envelope protein gp120. It is gp120 that binds to the CD4 receptor molecule of the T-helper lymphocytes which HIV attacks. Pert and her colleagues said that peptide T can block the receptor on the lymphocyte, preventing the virus from infecting these cells.

Other researchers working with HIV said soon after Pert published her research that they could not reproduce her results. Some scientists were also worried that one initial study of peptide T in patients had already taken place and another was planned. One researcher said: 'It is unusual for somebody to start a clinical trial based on somewhat surprising results from one laboratory that have not been confirmed by other groups.' The substance may be harmful, he warned. In addition, research had proved that zidovudine, for example, was effective, and could help to keep patients with AIDS alive until better treatments came along.

One group of researchers reported their results in the *Lancet*. They said that even when they added peptide T in concentrations 10^3 times greater than those which, in Pert's experiments, had completely inhibited replication of HIV, the substance had no effect on viral replication. Nor did peptide T block the binding of the gp120 envelope protein to the CD4 receptor on the cells. These scientists added that the peptide used by Pert's team comes from a highly variable region of the viral protein gp120. In a study to identify the same region in fifteen different strains of HIV, not one of the eight amino acids remained the same in all strains.

Pert and her colleagues responded in a letter to the *Lancet* in September 1987. The part of gp120 which contains the sequence known as peptide T is similar in many different strains, they maintained. Although some amino acids change, their replacements are structurally and functionally similar. So the sequences remain analogous. Pert's group also pointed out that the tests used by the other researchers used high

148

concentrations of both virus and receptor – conditions which do not occur in the body. Furthermore, researchers in Pert's laboratory, as well as an independent group, have found that peptide T was less able to stop the virus infecting cells as the concentration of virus increased. The letter concluded: 'Failure to detect activity in assays of uncertain physiological relevance should not be grounds for postponement of clinical trials of peptide T . . .'

Pert, together with Lennart Wetterberg of the department of psychiatry at the Karolinska Institute in Stockholm, and other colleagues, had already published a letter in the Lancet in January 1987 saying that they had given peptide T to four patients with AIDS. They said that the numbers of lymphocytes in these patients increased, and none of the patients lost any weight.

In May 1987, a larger trial of peptide T, in thirty-six patients, began in Sweden. However, three of the original four patients who took peptide T are dead. A second trial of peptide T, involving twelve patients, also began in the US in 1987. Scientists hope that the results of these studies, when they become available, should settle once and for all whether peptide T is a useful treatment for AIDS.

Peptide T is not the only potential therapy to have spawned controversy. AL721 is another. This substance, a mixture of natural lipids (fats) made from egg yolks, is a yellow oily liquid, which can be taken spread on bread or in orange juice. The 'AL' stands for active lipid; the '721' represents the ratio of the three different lipids that the mixture contains. When the gay press began to publish recipes for similar preparations that individuals could make at home, many people with AIDS, impatient with what they saw as the medical establishment's failure to speed up investigations into novel treatments, eagerly grasped the opportunity to take matters into their own hands. Bizarre headlines – 'Eggs against death' was one – in gay newspapers symbolized a growing preoccupation with AL721. Many gay pressure groups in the US and Europe organized themselves to obtain and distribute bulk supplies of the necessary ingredients.

The tragedy of AL721 is that it may now prove impossible to evaluate scientifically. Once people perceive a treatment as useful and effective, it becomes very difficult for doctors to enrol patients into a randomized, placebo-controlled trial (see p. 42). Those who take part may be so concerned that they may not be receiving the active preparation that they may secretly take their own home-made version.

This kind of action invalidates the results and makes analysis impossible. The blame for this state of affairs does not lie with the people who decide to try to do something for themselves, however, for the progress of AL721 through the intricate path of scientific evaluation has been painfully slow.

AL721 was discovered by researchers at the Weizmann Institute of Science near Tel Aviv in Israel. They developed the substance because they found that it could remove molecules of cholesterol from cell membranes. Cholesterol builds up in cell membranes during old age, reducing the fluidity of the membrane. Doctors have given AL721 to elderly people in order to improve their immune systems through the drug's effect on cell membranes. Meir Shinitzky, a biophysicist at the Weizmann Institute, says that American researchers at the University of Virginia discovered in 1978 that some viruses need large amounts of cholesterol in their membranes if they are to infect cells successfully. The researchers could prevent the viruses from infecting cells if they removed the cholesterol; if they replaced it, the infectivity returned.

The first time that AL721 was mentioned as a promising candidate for the treatment of AIDS was in November 1985. A group of American researchers wrote a letter to the *New England Journal of Medicine*. AL721, they said, seemed to be able to inhibit HIV from infecting human lymphocytes. 'A major need exists,' they added, 'for a safe, effective agent that can be administered alone or in combination with other drugs for the treatment of AIDS. Since AL721 does not directly inhibit reverse transcriptase, it should not produce the undesirable side effects that are associated with reverse transcriptase inhibitors . . .' But much more work would be needed to find out how useful AL721 might be in treating patients. However, two years later, scientists were no wiser about the effectiveness of this substance.

No one is quite sure how AL721 is supposed to work. According to one theory, the membrane of HIV – which is derived from the infected cell as the virus buds out – contains a lot of cholesterol. Perhaps removing cholesterol from the membrane changes its density, making it much more fluid. Possibly, the envelope proteins that normally project from the surface of the virus sink down into the membrane, which then conceals the sites on the viral proteins that are important in binding to the T-helper cell.

Shinitzky, together with colleagues Yehuda Skornick and Zvi Bentwich, first used AL721 to treat patients with cancer. One man, with a type of cancer called lymphoma, had AIDS. Shinitzky told New Scientist in May 1987: 'AL721 had a remarkable effect on this patient.' The researchers then decided to treat AIDS patients with AL721, even though they did not have permission from Praxis Pharmaceuticals (which later became Ethigen Corp.), the company to whom the Weizmann Institute had sold the manufacturing rights for the substance. They gave AL721 to sixteen patients with AIDS. In all except one, who died, there was considerable improvement. They put on weight and had fewer opportunistic infections.

Initial tests have also taken place at St Luke's Roosevelt Hospital in New York. In the summer of 1986, doctors there gave AL721 to eight people infected with HIV. These patients all had persistently enlarged lymph glands. Arthur Englard, one of the doctors involved, told New Scientist in May 1987: 'Results were very good but you have to realize that a sample of eight people is not statistically significant.' In five patients, there was a 'dramatic decrease' in the amount of virus found in their blood. In some, there was no evidence of viral activity – as indicated by levels of the viral enzyme reverse transcriptase – after two months on AL721. The patient's immune systems improved, and there seemed to be no side effects to the treatment.

At the end of the trial, when they stopped taking AL721, the virus reappeared in the blood of all eight patients. Englard and his colleagues, Michael Grieco and Michael Lange, then had difficulties in getting further quantities of AL721. It took them two months before they obtained fresh supplies.

Doctors and researchers around the world have had plans for a large trial of AL721 since about the beginning of 1986. Hospitals in seven cities, including London, Tel Aviv and New York, were to have taken part in a trial that would involve 300–400 patients. In early 1987, doctors at St Mary's Hospital in London said that they had been waiting for delivery of the preparation for about eighteen months. Six months later, Englard said that his group was very frustrated with the hold-up on AL721. The Food and Drug Administration, which controls the testing and licensing of new drugs in the US, had approved a study to determine the toxicity of AL721. Considering how many people had already taken AL721, Englard said, 'That's not the study to do at this time. It's just a waste.' The results of the study at St Luke's Roosevelt Hospital had been available for over a year, yet the research was no further forward. 'It infuriates us,' Englard said.

Meanwhile, HIV-infected people around the world have been mixing lipids in their own kitchens in the hope of keeping AIDS at bay. They may have been wasting their time and resources. Or they may have taking an effective preparation. Lack of proper scientific evaluation means that no one can tell. If AL721 does do some good, it seems scandalous that neither doctors nor AIDS patients should, by 1988, not know for certain.

AL721, if it works, probably acts by preventing the virus from binding to its target cell. Once inside the cell, HIV discards its protein coat. Substances that prevent the process of uncoating could have a role to play here. The loss of the coat releases the genetic material – RNA – and the enzyme reverse transcriptase into the cytoplasm of the cell. It is at this point that the drugs that inhibit reverse transcriptase, mentioned earlier in this chapter, come into their own. In the absence of such inhibitors, however, the cell makes viral DNA from the virus's original genetic material, its RNA. Once the DNA has integrated itself into the genetic material of the human cell, it begins to direct the manufacture of viral proteins and more copies of viral RNA, ready for the assembly of new viruses. The complex process of making new viruses provides many more targets for novel antiviral drugs.

The drug ribavirin, for example, is said to act by selectively preventing the virus from completing copies of the genetic material RNA which carries the information necessary for the manufacture of viral proteins. Ribavirin (trade name Virazole), which is made by the Californian company ICN Pharmaceuticals, is already licensed in the US for the treatment of respiratory infections, caused by respiratory syncytial virus, in children. It is also effective in the treatment of viral diseases such as Lassa fever. Yet tests of its activity in the treatment of HIV infection had not, by 1988, provided clear-cut results.

At the Third International Conference on AIDS held in June 1987, Peter Mansell, a doctor at the University of Texas in Houston, presented the results of a study of 163 men, some of whom were treated with ribavirin. All the patients had persistently enlarged lymph nodes. The trial divided the patients into three groups: one group took 800 mg ribavirin a day, a second group took 600 mg a day and the third set were given a placebo. During the period of the trial, which took twenty-four weeks, ten of the fifty-six patients taking placebo developed AIDS. In the group taking 600 mg ribavirin a day, six men developed AIDS. None of the patients taking 800 mg a day progressed to AIDS, however.

Some members of the audience at the conference criticized Mansell's presentation. Frank Young, commissioner of the US's Food and Drug Administration, in an unprecedented breach of tradition at a scientific meeting, told the session that although the FDA had examined the study in detail, it had 'no evidence of effectiveness'. The reason, he said, was related to the patients' levels of T-helper cells. Taken on average, the numbers of T-helper cells in both treatment groups

153

and the placebo group were similar. But there were seven patients with fewer than twenty T-helper cells per millilitre of their blood – a very low count indeed. Six of these seven were in the placebo group, which could, therefore, have been expected to do badly. The next speaker asked Mansell whether he could consider the results significant when about 18 per cent of patients in the placebo group had developed AIDS within six months – an extraordinarily high percentage.

The previous month, Young had made a similar statement about ribavirin to a congressional hearing. The same committee heard from a doctor based in New York, Bernard Bihari, that officials from ICN Pharmacuticals, the manufacturers of ribavirin, had approached him. They said that ribavirin was about to be licensed for the treatment of a respiratory condition in children; once it was, they would be able to make the drug available to him and other physicians in private practice for the treatment of AIDS – even though it had no licence for this use.

Later in 1987, ICN Pharmaceuticals resubmitted their data to the Food and Drug Administration after further analysis. Several independent scientists and statisticians had re-evaluated the results of the study. Many of them said that they believed that ICN should be allowed to carry out more studies of ribavirin. Contrary to the view of the Food and Drug Administration, these scientists said that patients in the study had been randomly allocated to treatment and placebo groups in an acceptable way. In September 1987 the Food and Drug Administration agreed to allow the company to conduct further trials with ribavirin.

One anomaly in the story of ribavirin is that there is an organized black market in the drug over the border between California and Mexico. Gay pressure groups in California have provided detailed advice on how to obtain ribavirin and other drugs that are not available in the US but can be bought in Mexico. The safest strategy on returning through customs, says the advice, is to declare a personal supply of the drug. It is legal for people to bring in to the US enough ribavirin for about one month's use.

As with AL721, it seems unfortunate that many people have spent their money on illicitly buying a drug the effectiveness of which was still unclear at the beginning of 1988. An American pressure group called the National Gay Rights Advocates (see p. 157) has pointed out that in these circumstances, 'a person's use of the drug [ribavirin] is totally unsupervised, without any benefit to research in this country'. By the end of 1987, enough people had probably taken enough ribavirin to answer the question of whether this drug is effective against HIV many times over. Because these patients were not taking part in clinical trials, however, much potentially useful information has been lost.

Several vulnerable points in the virus's life cycle still remain for drug designers to exploit. Molecules of RNA carry the information contained in the viral DNA out into the cytoplasm of the cell where proteins are made. In theory, it might be possible to block this messenger RNA with short chunks of RNA which present a complementary sequence of information. One problem with this approach will be how to deliver the alien RNA to the cell.

Another weakness in the virus's defences occurs at the time when the cell is producing viral proteins. The virus manufactures several of its proteins in large chunks which depend on viral enzymes, called proteases, to cleave them into smaller functional segments. Substances that could inhibit these enzymes would probably prevent the assembly of new viruses. Investigators in Britain here already worked out a model for the structure of one of these enzymes, which they say will be useful when it comes to designing inhibitors of these molecules.

Researchers have also identified a small viral protein which boosts the rate of manufacture of other viral proteins. Any chemical which could inhibit this regulatory protein would greatly reduce the number of viruses produced by a single infected cell.

One drug that may act by interfering with protein synthesis is fusidic acid (Fucidin). Fusidic acid is an antibiotic that was discovered in the 1960s. Doctors in Copenhagen and London decided to investigate this drug for the treatment of AIDS after a patient with AIDS who had received it made a dramatic recovery. The patient, a fifty-eight-year-old man, had tuberculosis and several other infections. His doctor decided to add fusidic acid to the mixture of drugs prescribed for the patient because it is known to act against the bacterium that causes tuberculosis and to boost the actions of other antibiotics.

The man recovered: his fever went, he put on weight and he went back to work. The researchers decided to investigate further in order to find out whether his improvement was due to the antibacterial properties of the drug, or to some direct effect on the virus. To their surprise, they found that fusidic acid completely inhibited the infectivity of the virus, as measured by standard tests. The dose of fusidic acid needed for this effect was relatively high, but it is possible to achieve these concentrations in a patient's blood.

Fusidic acid does not work by inhibiting reverse transcriptase. Scientists have suggested that the drug may inhibit the synthesis of viral proteins, just as it prevents the manufacture of bacterial proteins. There is some evidence that fusidic acid disrupts the normal association between messenger RNA and cellular structures called ribosomes, during a critical step in protein synthesis. Alternatively, it may inhibit the viral enzymes that break up the large precursor proteins of the virus into their smaller components.

Trials of the effectiveness of fusidic acid in patients with AIDS began in Copenhagen and London in late 1987. If it is shown to be effective, fusidic acid will have several advantages over other candidate drugs. It can be taken by mouth for long periods, it can penetrate the central nervous system and it has few side effects. It is also relatively cheap.

Scientists have yet another potential strategy available to them in their search for effective drugs to treat AIDS. They may be able to find substances that can disrupt the assembly of new viruses. One candidate for this role is a substance called castanospermine, which is extracted from the seeds of an Australian tree, the Moreton Bay chestnut. The proteins in the envelope of the virus are densely covered with sugars. Castanospermine seems to prevent the virus from adding these sugars to the proteins in the normal way.

Studies have shown that castanospermine can inhibit replication of the virus in cell cultures in the laboratory. Scientists are not sure how this drug works, however. It does not prevent the virus from binding to the cell, but it may stop the virus from entering. Alternatively, by interfering with the processing of the viral envelope proteins, castanospermine may prevent the virus from leaving the infected cell.

Castanospermine may be too toxic to use in patients. It may be useful, however, in conjunction with other drugs. Furthermore, it may be possible to design new drugs that act in the same way by analysing the action of castanospermine. Scientists have already begun to test several similar drugs.

Many people infected with HIV, especially those who have already developed AIDS, believe that the standard procedures for testing novel therapies are unnecessarily slow. If a disease inevitably kills within three years, sufferers are not going to be able to wait four years for new drugs to be tested. The families and friends of people with HIV infection, as well as the sufferers themselves, are understandably angry and frustrated at the lengthy machinations of the regulatory bodies that control the approval of new drugs.

In the US, even changes in the Food and Drug Administration's regulations on new drugs for 'desperately ill' patients, in force since June 1987, do not seem to have improved matters. People have resorted to the courts to try to find out why there have been undeniably long delays in evaluating certain treatments. One pressure group, the National Gay Rights

Advocates (NGRA), has tried to sue the heads of several American government departments for failing to act swiftly enough in developing new treatments. NGRA alleged that the Food and Drug Administration has accelerated testing and approval only for drugs developed or sponsored by the National Institutes of Health (NIH), a government body.

The lawsuit claims: 'NIH concentrated its research into NIH-sponsored drugs, or into drugs developed by companies with which NIH or its researchers had developed special relationships, such as Burroughs Wellcome or Hoffman-LaRoche . . . NIH ignored or seriously delayed consideration and testing of other promising drugs.' The Food and Drug Administration, the group alleges, is applying more stringent procedures and requirements to drugs developed privately than it did to zidovudine. 'NIH's decision,' the lawsuit continues, 'were affected by essential conflicts of interest, namely, royalty payments from manufacturers licensed to develop NIH-sponsored drugs.'

Public loss of confidence in the system in the US has already led to one state taking matters into its own hands. On 28 September 1987, the governor of California, George Deukmejian, signed an emergency bill to allow California to carry out its own testing and licensing of drugs. It would have been difficult for him to refuse, for the bill had wide support from politicians. The state Legislative Assembly had passed it by seventy-nine votes to nil, and the Senate by thirty-eight to nil. At the time, the state attorney-general, John Van de Kamp, said: 'This bill is the state of California's announcement that, in the face of an extraordinary medical emergency, business as usual just isn't good enough.'

The new law means that California can test, manufacture and distribute experimental drugs within the state – provided that all raw materials come from within California. Some commentators believe that the result could be a black market in novel drugs over the border with California. Dubious manufacturers and suspect therapies could mushroom. A spokesman for the Food and Drug Administration, which denies that it is being too slow in testing new substances, warned that there may be dangers in approving new drugs too quickly. The state of Nevada, for example, licensed a compound called laetrile. People used this substance, which is highly toxic, to treat cancer, even though there is no evidence that it is effective.

Even bona fide drugs companies in California that decide to take advantage of the new law will have their problems. For a start, by capitalizing on their new-found freedom, they may risk alienating the Food and Drug Administration. This could make it difficult for them to have new products approved throughout the US. In addition, even though there is a potentially huge market for drugs against AIDS in California, companies may still not find it worth their while to invest funds in the testing of new drugs for sale in one state alone.

Few people are suggesting that scientific evaluation of drugs for AIDS should be discarded. No one would gain if it were. The history of medicine holds plenty of examples of treatments that have turned out to do more harm than good. Without proper assessment, hundreds of remedies of doubtful usefulness would flood into the hands of patients desperate to try anything, as long as it gave them hope.

That said, many individuals believe that testing is taking too long, particularly for some drugs made by small private companies. Whether this is the fault of the companies in not providing appropriate data, or whether the regulatory bodies are to blame, is not clear. What is certain is that, particularly in the US, many people are taking drugs bought on the black market, or made up in their own homes. They are spending both time and money on therapies which may not work. It is time that they knew one way or the other whether they are doing themselves harm or good.

CHAPTER 11

TOWARDS A VACCINE

One of the most tantalizing features of the human immuno-deficiency virus is that it invades the very cells that should be capable of fighting it off. Furthermore, in contrast to the response to most other infections, most of the antibodies that the body produces against the virus are powerless against it. These characteristics of HIV infection make it peculiarly difficult for scientists to develop a vaccine.

For a start, scientists have no evidence that it is possible to produce immunity to HIV. Infection with a microorganism normally stimulates the immune system to produce antibodies and specialized cells which eliminate the intruder. Provided that the person recovers, he or she is then immune from further attacks by the same microbe. The aim of a vaccine is to mimic the natural stimulation of the immune system by the bacterium or virus in question, without the person having to suffer the disease. In HIV infection, although the body produces some antibodies capable of inactivating the virus, these are present only at very low levels. Instead of recovering, individuals suffer persistent infection. The antibodies in their blood are a sign that the virus is still present, and do not indicate immunity. Doctors have found no one who appears to have recovered from the infection with immunity to further attacks.

Another obstacle to making a vaccine is that humans are

the only animals susceptible to HIV. Normally, it is possible to find an animal which can develop the disease against which the vaccine is required. Using such an 'animal model', scientists can then test whether their potential vaccine works by trying deliberately to infect the vaccinated animal with the microorganism that causes the disease. If the vaccine protects the animal against this 'challenge', there is a good chance that it may work in humans, too.

Although it is possible to infect chimpanzees with HIV, so that they produce antibodies, none has so far developed the disease. Because of this, many researchers are uncertain about the value of chimpanzees in testing a vaccine. Given that chimpanzees are an endangered species in short supply, as well as the general objections to conducting these tests on such intelligent primates, this approach has its limitations.

The lack of an immediately obvious animal model may not hamper research into a vaccine too greatly, however. Scientists may be able to overcome the problem by studying similar diseases in other species. For example, cats and monkeys both suffer infections caused by retroviruses, with symptoms of immune deficiency. Researchers in California reported in February 1987 that they had discovered a new retrovirus in cats with a chronic AIDS-like condition.

Some species of monkey can also suffer from a form of AIDS called SAIDS. The initial 'S' stands for simian, from the Latin for ape. The agent that causes SAIDS is called simian immunodeficiency virus (SIV). This virus is related to the human immunodeficiency virus. The simian virus causes SAIDS in macaques in captivity, but this species does not appear to carry the virus in the wild. Other African monkeys, particularly the African green monkey, commonly have antibodies to SIV. But SIV does not appear to cause disease in this species.

By studying the biology of retroviruses that cause AIDS-like disease in other species, some scientists believe that they will discover information valuable to the fight against AIDS in humans. If they are able to develop vaccines against retroviruses in other animals, they may be able to use the same technology for a vaccine against AIDS in people.

Estimates of how long it will take to generate and test an effective vaccine vary. Trials of candidate vaccines began on human volunteers in the US in 1987. With this in mind, some optimists say that a vaccine may be available by 1989. In 1987, other scientists believed, perhaps more realistically, that it could be another ten years or longer before a vaccine could be available to the general public.

The prospects of a widely available vaccine may be distant, but researchers have not allowed this to deter them from exploring other ways of manipulating the immune system. Traditionally, vaccines are used to protect people from diseases that they have not yet encountered. AIDS is rather different from most diseases that humans suffer, however. Its incubation period is long and there is some evidence that the initial phase, during which there are no symptoms, may be the result of an immune reaction that manages to fight off the infection for a limited period. So it may be possible to find ways of manipulating the immune systems of people who are already infected with HIV to prevent them from developing AIDS itself. Vaccines designed to protect healthy uninfected people may have this effect. Doctors and scientists have also investigated ways of compensating for the abnormalities of the immune system in AIDS. One idea was to transplant bone marrow, which manufactures T-cells, in an attempt to boost the number of T-cells. Apart from the obvious difficulty of finding suitable donors, such a strategy seems to be fundamentally flawed: the virus infects the transplanted cells, too.

Another tactic is to administer not cells of the immune system but the molecules that these cells normally produce. White blood cells of the T-helper variety produce these substances, called cytokines, when they are stimulated by an antigen. The cytokines influence the behavior of other cells, expecially macrophages, the large mobile cells that have the task of trapping, engulfing and eliminating microorganisms. There are several kinds of cytokine, with varying functions. Some attract macrophages. Others inhibit macrophages from leaving once they have reached the site of infection. Other cytokines, the interferons, activate macrophages, so that these cells are better able to kill microorganisms. Interferons can also enhance the activity of other cells of the immune system.

Researchers postulated that if T-helper cells were in short supply in people with AIDS, the immune defects in these patients might be due to failure of these cells to carry out their normal role. Perhaps it would be possible to compensate for this deficiency by giving the person doses of cytokines. A second strategy would be to boost the production of T-helper cells in some way.

Several teams of doctors and scientists are working on these approaches. Researchers in San Francisco decided to try treating patients with a cytokine called interleukin-2. In healthy people, T-helper cells that have been activated by an antigen release this substance, which stimulates other cells that play an important role in the immune response. The T-helper cells of patients with AIDS, however, produce reduced quantities of interleukin-2. Laboratory tests suggested that adding interleukin-2 might be helpful, so the San Francisco group gave this substance to eighty-seven patients with AIDS in 1986. The results, they said, were disappointing. None of the patients showed any improvement in either their immune systems or their general condition.

Research into other cytokines continues. There has been much interest in the role of the interferons. Interferons have a variety of functions, including helping to eliminate tumour cells as well as viruses and other microorganisms. Laboratory tests have shown that some types of interferon can inhibit the replication of HIV in blood cells. Scientists believe that these substances act by preventing the normal assembly and release of viruses from infected cells. Tests of interferons in patients with AIDS have produced inconclusive results, however. Some doctors have reported successfully using interferons to treat Kaposi's sarcoma; others say that it can be helpful in AIDS patients suffering from allergic conditions, such as asthma. However, most studies have been on very small groups of patients, which makes it difficult to draw firm conclusions about the results.

Paradoxically, some of the symptoms experienced by patients with AIDS may be due to exceptionally high levels of interferons. When doctors used interferons to treat other viral diseases, they noticed that patients suffered symptoms such as fever, chills and diarrhoea. This suggests to some researchers that, in AIDS, the cells of the immune system may release inappropriately large amounts of some cytokines, such as certain interferons. In support of this theory, other doctors have found that patients who survive longer and have fewer opportunistic infections tend to have lower levels of a certain type of interferon than patients who succumb to AIDS at an earlier stage.

Other strategies involve giving patients substances that stimulate the production of cytokines. One treatment, called ampligen, is said to work in this way. Ampligen consists of special sequences of the genetic material RNA. Although RNA normally has only one strand of smaller molecules, in ampligen the RNA is double-stranded: it has two strands wrapped round each other. Twenty-two doctors and scientists wrote to the *Lancet* in June 1987 to report their experience of giving ampligen to ten patients with AIDS or other symptoms related to HIV infection.

Researchers initially tried using double-stranded RNAs to treat patients with cancer more than ten years ago. Double-stranded RNAs are known to promote the production of various cytokines. In addition, several enzymes which probably assist interferons in their actions against viruses also need double-stranded RNAs in order to operate. The theory was that it might be possible to harness the body's own defences against tumour cells. In these first trials, the patients suffered such severe side effects that doctors had to abandon this approach. Researchers then found that by modifying the RNA – the result was ampligen – they could eliminate the toxic effects. They later discovered that ampligen could inhibit replication of HIV in cells grown in the laboratory.

The next step was to see what effect ampligen had on patients with HIV infection.

The researchers found that ampligen seemed to remove the genetic material of HIV from the patients' blood. The patients felt better. Their levels of T-helper cells either stayed the same or increased. Patients with symptoms related to HIV infection, rather than AIDS itself, seemed to respond best to the treatment. Although the results of this study sounded promising, the researchers warned that it was not possible to conclude whether patients given ampligen might live longer. More studies would be needed, they said. They planned to investigate ways of usefully combining ampligen with antiviral therapies such as zidovudine. There is some evidence to suggest, for example, that ampligen may act with zidovudine against the virus, which may allow doctors to reduce the dose of zidovudine. Unfortunately, both drugs are extremely expensive.

There are other ways of compensating for some of the deficiencies of the immune system in people with AIDS. One method is to boost the number of T-helper cells with substances naturally found in the body. Researchers have already had good results using a substance with the cumbersome name of 'granulocyte-macrophage colony-stimulating factor', or GM-CSF for short. Jerome Groopman, from the New England Deaconness Hospital in Boston, Massachusetts, and his colleagues in the US, reported in the *New England Journal of Medicine* in September 1987 that GM-CSF could increase the numbers of white cells circulating in the blood. To judge from laboratory experiments, GM-CSF has a multitude of effects on the immune system. It stimulates the immature cells of the bone marrow – which produces all types of blood cells – to divide and develop. It also influences the mature cells of the immune system. For example, it boosts the ability of macrophages to kill tumour cells. Researchers have also shown that GM-CSF can inhibit the replication of HIV in infected cells grown in the laboratory.

But does the same thing happen in patients? Groopman's team gave GM-CSF to sixteen patients with AIDS. Each patient first received an injection, followed, two days later, by an intravenous infusion of GM-CSF which lasted two weeks. Within six hours of the initial injection, the numbers of white cells in the patients' blood rose. Groopman and his colleagues suggested that this elevation may have been due to release of mature cells from the bone marrow. But the number of white cells continued to rise during the fourteen days of infusion, suggesting that GM-CSF was able to stimulate the bone marrow to produce more cells. Levels of white cells returned to their previous levels, however, shortly after the infusion came to an end.

There is a persistent problem with studies of this kind. No one knows what the relationship is between the number and function of T-helper cells and how ill patients with AIDS feel. Further studies should help to solve this problem. In the meantime, substances such as GM-CSF may help patients who cannot tolerate drugs such as zidovudine because of the effect of these drugs on the bone marrow. GM-CSF may also prove useful for people with immunosuppression caused by conditions other than AIDS, such as those receiving radiation therapy or taking toxic drugs for the treatment of cancer. They may be able to tolerate larger, more effective doses without suffering the ill effects of damage to the bone marrow.

Many companies and researchers are racing to develop other therapies that would act on the immune systems of infected people. Small trials of drugs dubbed 'immune regulators', including, for example, Imreg-1, Immuthiol and thymostimulin are continuing both in the US and in Europe. By the end of 1987, however, most studies on the effectiveness of these substances were small and the results generally inconclusive.

Many of these potential treatments involve newly discovered substances, often prepared or manufactured with the benefit of the latest techniques for manipulating genetic material such as DNA or RNA. Traditional methods of boosting the immune system are not being neglected, however. Researchers are trying to find out whether it is possible to treat patients with AIDS by giving them antibodies from other individuals in the early stages of infection. Perhaps, scientists have postulated, these antibodies might allow the person to revert to the early phase of the infection, when there are no symptoms.

Teams in Cambridge and London are investigating this approach. They are giving AIDS patients plasma taken from healthy people who have high levels of antibodies to HIV. (Plasma is the fluid part of the blood that remains after the cells have been removed.) By pooling plasma from many different donors, the researchers hope to offer the recipients a better chance of receiving effective antibodies.

Doctors in the US are relying on a related strategy in an attempt to protect children with AIDS against common infectious diseases. Children with AIDS have the problem that they have no immunity to the usual childhood diseases. In addition, the defects in their immune systems seem to interfere with their ability to make antibodies in response to new infections. So doctors are trying to find out if giving a substance called immunoglobulin – essentially a mixture of antibodies – can confer immunity. Treatment with these antibodies, prepared from blood donated by many people, is called passive immunization. Some researchers believe that this therapy cuts the number of infections that develop in affected children. Critics of the treatment, however, say that it is too expensive and that the available results do not provide convincing proof that the method works.

Passive immunization is always second-best to active immunization – the kind that a vaccine aims to induce by stimulating the body to produce its own antibodies. Passive immunity lasts only as long as the immunoglobulins that are transfused. Active immunity, on the other hand, confers long-lasting protection against the infectious agent. The individual's immune system also retains a memory of the foreign antigen, so that the immune system responds to subsequent exposure to the bacterium or virus more rapidly and effectively.

Vaccines are traditionally used to protect people before

they encounter a particular disease. Theoretically, however, a vaccine against HIV given early in the infection may help to prevent the development of AIDS. It seems logical that, if it is possible to stimulate an effective immune response at all, this response might also be able to protect those already exposed to the virus. Studies to test this hypothesis began in 1987.

The man behind these tests is Jonas Salk, head of the Salk Institute in La Jolla, California. It was Salk whose pioneering work led to the development of a vaccine against polio-myelitis in 1954. This vaccine, like the preparation developed against AIDS, also consists of whole viruses that have been inactivated.

Salk refuses to call the product a vaccine; he prefers the term 'potential immunotherapeutic agent'. He is collaborating with the Immune Response Corporation, a company also based in La Jolla, which will produce quantities of irradiated virus for the trials. The virus is exposed to radiation in order to inactivate it, so that it is no longer able to infect cells. There are several historical precedents for a vaccine based on whole inactivated viruses being effective. One is polio vaccine. Another is a vaccine that researchers at the University of California at Davis developed several years ago, which seems to protect monkeys against simian AIDS.

Doctors in Sacramento appealed in late 1987 for forty volunteers to take part in the trial. They wanted people to take part who were infected by HIV but who showed no symptoms. Only half of them would receive the new agent. Doctors would then follow up the forty individuals to see whether people in the group which receives the therapy take longer to develop AIDS than those who receive the placebo.

By giving this preparation to people who are already infected with HIV, Salk and his colleagues are circumventing the ethical problems that would arise if they wanted to test the agent on uninfected individuals. For example, although the irradiated viruses are unable to infect cells, doubts would still remain about the advisability of giving a preparation that contained viral genetic material to uninfected people.

Traditionally in the manufacture of vaccines, the way to avoid the risks associated with whole microorganisms – even if they are inactivated – has been to identify which component of the microorganism stimulates effective immunity. By isolating the element and giving it alone as a vaccine, it is often possible to confer protection against the whole virus or bacterium concerned. The problem with HIV is that, particularly as there is no natural immunity to the infection in humans, no one is sure which part of the virus will provoke the immune system in an appropriate way. A further difficulty, as with many viruses, is that HIV mutates very rapidly – many times faster than the influenza virus. HIV can even mutate in a single individual as time passes: the virus isolated from a person today will not be identical to the virus isolated from the same person in a year's time. To be effective, a vaccine would have to protect against all isolates (strains) of HIV.

Viruses are not alive in the strict sense of the word. So, technically speaking, it is not possible to talk about killing them. Instead, scientists use the word 'neutralize' to describe the process by which the immune system inactivates and eliminates viruses. Antibodies that can have this effect are called neutralizing antibodies. Someone who is infected with HIV has many antibodies against different components of the virus. In other words, each viral protein acts as an antigen against which the immune system makes antibodies. In fact, there are even antibodies against individual sites on each protein. It may turn out that neutralizing antibodies, which infected people are unable to produce in large amounts, recognize a particular site on a single protein. By binding to that site, these antibodies may knock out the whole virus. The problem is, how do scientists identify the crucial component?

Researchers began their search for a vaccine by screening each protein of the virus to see whether these molecules could, by themselves, induce neutralizing antibodies against HIV. They found that if they injected purified envelope proteins into experimental animals, these proteins did indeed induce neutralizing antibodies. But there was a drawback. In laboratory tests, the antibodies could protect susceptible cells against infection – but only against the original isolate of the virus. Rapid mutation of the envelope protein meant that antibodies to the original strain failed to recognize other isolates.

It is not possible for the whole envelope protein to mutate. Some part of it must remain the same if the molecule is to retain its function of binding to human T-helper cells. Scientists call these unchanging areas of the protein 'conserved' regions. Antibodies to the conserved regions might be able to bind to all isolates of the virus. But there is a catch here, too. These antibodies may well not be neutralizing antibodies. On the other hand, they may still be able to block the attachment of the virus to the T-cell (see p. 180).

Yet another complication is that antibodies may not be the only answer to immunity against HIV. The cells of the immune system may be important, too. Unfortunately, scientists know less about the role of the cells of the immune system – 'cell-mediated immunity', as it is called – than about the role of antibodies. Some researchers have speculated that humans may not be able to involve the cells of the immune system in a response to HIV. However, some initial tests of vaccines suggest that they may be wrong (see p. 173). One reason why cell-mediated immunity may be particularly important is that HIV probably passes from one person to another by the transmission of infected cells, as well as free virus. This suggests to some investigators that a vaccine which stimulates the production of antibodies alone may not be enough. It may be necessary to involve specialized immune cells as well.

It may be years before scientists establish the importance of these factors. Yet researchers are pressing ahead with tests on candidate vaccines, for there is no time to lose. In the US, the Food and Drug Administration has already authorized trials in humans of a vaccine based on viral envelope protein. In HIV, the envelope protein is in two parts: gp120, which sticks out from the viral membrane, and gp41, which spans the membrane. These two proteins start out as one large mol-

study also found that lymphadenopathy did not seem to correlate with a high risk of AIDS.

The researchers concluded:

> The incidence of AIDS after herpes zoster is approximately linear, with half of the AIDS cases arising within four years of zoster. Extrapolation of this linear trend suggests that eight years would be about the longest incubation period between zoster and AIDS. If we add to this another two to seven years between HIV seroconversion [development of antibodies to HIV] and zoster, the risk of AIDS developing after HIV seroconversion must continue for at least ten to fifteen years.

Some patients infected with HIV first visit their doctors not because they are suffering from typical infections, but because they are becoming forgetful or apathetic or are unable to concentrate or initiate and control voluntary movements. Changes in behavior may also occur. These neurological symptoms are caused by HIV attacking the brain and nervous system. About 60 per cent of HIV-infected people have symptoms of this kind, while up to 90 per cent have abnormalities of the brain and nervous system that doctors can detect at a postmortem examination. Full dementia, similar to that which occurs in Alzheimer's disease, develops in about 80 per cent of patients with neurological symptoms within a year, according to some studies. Incontinence, inability to coordinate movements and slight paralysis are common. AIDS dementia, which is also called AIDS encephalopathy, is included in the new definition of AIDS which came into effect in September 1987.

When doctors first began to recognize that many patients with AIDS also developed mental disturbances, they initially thought that these symptoms might be due to the brain infections and tumours that some people with AIDS develop, or to psychological effects. Yet research has shown that HIV itself is to blame, even though it does not commonly infect nerve cells. Many studies have now confirmed that, in many AIDS patients, it is possible to isolate HIV and viral components from the brain and the cerebrospinal fluid, which bathes the tissues of the central nervous system. In some patients, the amount of virus isolated from the brain and cerebrospinal fluid is far greater than that present in the blood.

Postmortem examinations of the brains of people who had suffered from AIDS dementia showed that the tissue of the brain had shrunk. Under the microscope, researchers saw characteristic abnormalities. In some individuals, small groups of inflammatory cells appeared throughout the brain. Other patients had no inflammation, but spaces had appeared in some parts of the brain. Researchers also often found abnormally large cells with many nuclei which they called 'multinucleated giant cells'. These cells seem to resemble the white blood cells in laboratory cultures that fuse together when infected with HIV (see p. 101). Before long, several teams of researchers had identified the infected cells in the brain as monocytes and macrophages. Other workers also suggested that the 'multinucleated giant cells' probably started out as macrophages, too, because of similarities in their structure. David Ho, of the Cedars-Sinai Medical Center in Los Angeles, and his colleagues Roger Pomerantz and Joan Kaplan, put forward a theory to explain how HIV manages to gain access to the central nervous system. First, HIV infects the monocyte in the bloodstream. The infected cell then passes across the blood–brain barrier – the tightly linked cells that keep the bloodstream separate from the fluid that bathes the central nervous system – and into the brain. These researchers then suggest two possibilities. First, the infected cells might release toxic chemicals. These substances might damage the nerve cells and the glial cells, which protect the nerve cells and manufacture the insulation around their fibres. Or, these toxins might attract inflammatory cells which carry out the damage themselves.

The second possibility is that the infected monocytes might affect the cells of the blood–brain barrier, so altering the permeability of the barrier. Such a change would alter the delicately balanced environment of the central nervous system, upsetting the function of the nerve cells.

Another theory is that HIV may be able to infect glial cells. Some researchers have found that HIV can attack these cells under laboratory conditions. In addition, investigators have found evidence that HIV can, on rare occasions, infect nerve cells. But it is still not clear to what extent such infection accounts for the malfunctions of the nervous system in people with AIDS. Figure 11 summarizes some of the current theories about how the virus damages the brain.

Yet another proposal is that viral proteins released by infected monocytes could interfere directly with the function of nerve cells. Work by Ho, this time with other colleagues, provides support for this theory. They showed that part of the envelope glycoprotein of HIV is similar to a natural chemical called neuroleukin. Neuroleukin, a protein which is found in human skeletal muscles, brain and bone marrow, prolongs the life of embryonic nerve cells in laboratory cultures.

The researchers found that, in the absence of neuroleukin, 90 per cent of embryonic nerve cells died within forty-eight hours in a laboratory culture. If they added neuroleukin, however, up to 45 per cent of cells began to grow and develop within twelve hours. The same thing happened if they added a substance called nerve growth factor.

The team then tried adding HIV to cultures of nerve cells supported by either neuroleukin or nerve growth factor. They found that HIV consistently suppressed the growth of the nerve cells supported by neuroleukin, but did not affect those cells grown in nerve growth factor. After further tests, they concluded that the viral envelope protein, gp120, was able to inhibit the activity of neuroleukin, but not that of nerve growth factor. Subsequent analysis of the sequences of the amino acids, the small molecules that make up proteins, showed that neuroleukin and gp120 shared the same amino acids at fourteen out of forty-seven sites – a similarity of about 30 per cent. This similarity, the researchers said, appears in all 'isolates' or strains of HIV in which scientists have studied the sequence of the gp120 protein. In support of this work,

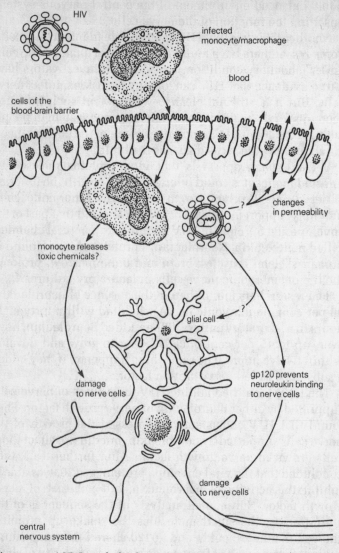

Figure 11. AIDS and the brain. Scientists are not sure which of these theories may account for the dementia and other neurological effects that some people with AIDS develop.

ecule, called gp160. It is gp160 that forms the basis of this vaccine.

A couple of decades ago, the available technology did not permit the preparation of quantities of purified viral proteins for a vaccine. Now, the techniques of genetic engineering, which allow scientists to manipulate molecules such as DNA and RNA, mean that it is relatively easy to manufacture these substances in bulk. The company making the vaccine, Micro-GeneSys of West Haven, Connecticut, is growing the gp160 in insect cells, making use of these new techniques. The first step in this process involves a virus, called baculovirus, which normally infects insects. Researchers insert the gene from HIV that carries the information for gp160 into the baculovirus. They then infect insect cells in culture with the engineered baculovirus. These insect cells then make quantities of gp160, which scientists can extract and purify. To make the vaccine, the protein is 'flagged' with a substance called an adjuvant. An adjuvant is a preparation which enhances the immune response to any antigen injected with it.

Scientists from MicroGeneSys plan to carry out a trial in collaboration with doctors from the National Institute of Allergy and Infectious Diseases in Bethesda, Maryland. The team hopes to recruit eighty-one people. All of them have to have tests to establish that they are not infected with HIV, as well as counselling on how to avoid infection. Fifteen of the eighty-one will be heterosexual, but the rest will be homosexual. Two thirds of the volunteers will receive the vaccine, which is called VaxSyn HIV-1. The rest will be given a control antigen to measure their basic immune response.

This trial aims to establish the safety of the vaccine and to find out what type of immunity it induces. Previous research has shown that VaxSyn HIV-1 can produce high levels of antibodies to gp160, gp120 and gp41 in experimental animals. According to a report published in 1987 in the *Journal of the American Medical Association*, tests suggested that these antibodies might have been capable of neutralizing the virus. When researchers added serum from the injected animals to cultures of human T-cells, along with HIV, the virus did not infect the cells. The animals' antibodies may have prevented the virus from binding to the T-cells.

Perhaps surprisingly, the Food and Drug Administration did not require the company to try to infect the experimentally vaccinated animals with HIV before allowing trials on humans to go ahead. Franklin Volvovitz, president and chairman of MicroGeneSys, told *New Scientist* in August 1987: 'Possibly that challenge will be done prior to subsequent trials.' He expected to have results of the first trial by April 1988. If these were satisfactory, tests to establish whether the vaccine is effective would probably take place at least two years after that.

Plenty of other groups are also working on vaccines based on the envelope proteins. One popular approach has been to insert the gene for the envelope proteins into the vaccinia virus. (Vaccinia is the virus that produces immunity to smallpox.) Cells infected with such genetically altered vaccinia manufacture gp160, which appears on their membranes. Researchers in the US have vaccinated chimpanzees with this type of preparation. They reported in *Nature*, in August 1987, that the chimpanzees did make antibodies to HIV. The researchers then tried to infect these animals with HIV. Unfortunately, tests carried out later showed that virus was present in all the animals. This team, led by Shiu-Lok Hu and colleagues from a company called Oncogen in Seattle, Washington, was cautious about drawing too many conclusions from these results. The immune response may not be the same in humans as in chimpanzees, they warned.

They may be right. Earlier in 1987, the French researcher Daniel Zagury, from the Pierre and Marie Curie University in Paris, reported in *Nature* the preliminary results of vaccinating himself and several others. He also used a genetically altered vaccinia virus containing the gene for the envelope protein of HIV, from a strain called HTLV-3B. Zagury and his co-workers in France and Zaïre said that they wanted to develop a method of vaccination that did not simply induce the production of antibodies but which also involved the cells of the immune system — 'cell-mediated immunity'.

Zagury first inoculated himself with the recombinant vaccinia virus, with no unusual side effects. Tests two months later showed neutralizing antibodies against HTLV-3B, but not to a second strain, called HTLV-3RF. Zagury then looked for evidence of cell-mediated immunity. He found that both strains of the virus could cause immune cells from his blood to divide and proliferate, although HTLV-3RF stimulated them less effectively.

A small group of Zaïrean volunteers, without evidence of infection with HIV, also received the vaccine. Two of these people, Zagury found, developed significant levels of neutralizing antibodies against the strain of the virus called HTLV-3B.

In June 1987, at the Third International Conference on AIDS, the ballroom of the Washington Hilton was packed with delegates and journalists eager to hear Zagury's latest results. On the evening of the second day of the conference, the Frenchman stood up to give his presentation. Zagury spoke rapidly, in heavily accented English. Many delegates, scientists included, were hard put to understand what he said, let alone grasp the science in his talk. When Zagury tried to leave the platform, pandemonium broke out. Mobbed by camera crews and journalists, Zagury tried to break free. Photographers and reporters chased him along the corridors of the Hilton before he finally escaped from them.

It was left to previous speakers to interpret what Zagury had said. Dani Bolognesi, of Duke University Medical Center in Durham, North Carolina, said: 'I think he's taken a step that has eluded us for some time.'

Zagury had told the conference that he had followed the initial inoculation (as described in his original paper in *Nature*) with a booster injection. The booster consisted of 'disabled cells'. These cells had been removed from him and the other vaccinated individuals, infected with vaccinia containing the gene for gp160, inactivated and then injected back into the same individuals. This procedure, Zagury said, seemed to make it possible to induce antibodies which would neutralize many different strains of the virus. In a statement which Zagury left behind before he made his swift exit, he said that it would not be possible to boost immunity in this way on a large scale. He and his colleagues planned to search for a more practical way of delivering a booster.

Promising as Zagury's work is, a vaccine based on vaccinia has several drawbacks. One is that anyone already vaccinated against smallpox might not mount a strong enough response to a genetically altered vaccinia virus. Another is that doctors would not want to give such a vaccine to people who are already infected with HIV. Any extra strain on the immune systems of people already infected with HIV runs the risk of activating the virus, with possible progression to AIDS. But it would not be possible to run a mass vaccination programme where everyone first had to have a test to see whether they were infected. For a start, there would be ethical problems: not everyone would consent to a test. And in some countries where health services are very basic, there would not be enough money to pay for both testing and vaccination – and for the administrative framework that such a project would need.

Another objection, according to some researchers, is that vaccines that utilize live viruses may be dangerous to health. Apart from vaccinia, scientists have also investigated the potential role of another virus, called adenovirus, which causes colds and influenza. Some researchers are worried that viruses such as these, genetically engineered to carry genes from HIV, may insert their own genes into human genetic material. William Jarrett, head of the department of veterinary pathology at the University of Glasgow, said in 1987 that research on the possibility of using live viruses in a vaccine against HIV has 'gone cold'.

There are good reasons, however, to explain why researchers should choose viruses such as vaccinia as a vehicle for anti-gens from HIV. Many investigators believe that, in order to make an effective vaccine, it is important to 'present' viral proteins to the immune system in a particular way – as part of the membrane of a cell or another virus, for example. There are other ways of doing this without using viruses such as vaccinia, however.

One alternative is a fake virus. A group of researchers in Britain, in collaboration with a company located in Oxford, called British Biotechnology, has managed to create a 'pseudovirus'. By manipulating the DNA of yeast cells and viral DNA, the scientists managed to combine proteins from yeast cells with parts of the envelope protein of HIV. When the researchers, led by Alan Kingsman and Keith Gull, looked at the result under the electron microscope, they saw virus-like particles. They say that experiments have shown that these fake viruses are 'completely harmless' and are able to stimulate the immune system to produce antibodies capable of destroying HIV. The method may be particularly useful for surveying segments of protein for their ability to induce the production of protective antibodies.

Another option is called the ISCOM. The name stands for immunostimulatory complex. Jarrett and his co-workers in Glasgow have already developed an effective vaccine for a viral disease in cats using this method. They are now trying to put the same technology to work for a vaccine against HIV.

ISCOMs are small, cage-like structures which make it possible to present viral proteins in an array, as they appear on the membrane of the virus (see figure 15). The viral protein which Jarrett's team is using is gp120. The researchers are making this in the same way as MicroGeneSys is manufacturing gp160 for its vaccine – in insect cells infected with a virus. The next step in making the ISCOMs is to mix the viral protein with a substance called Quil A, which comes from the bark of the Amazonian oak, in the presence of a detergent. In these circumstances, ISCOMs form naturally.

Jarrett says that tests in apes and monkeys over two years have shown that ISCOMs are safe. The theory behind his approach is as follows. People infected with HIV produce low levels of neutralizing antibodies that are probably ineffective. ISCOMs, however, are very good at presenting antigens such as viral proteins to the immune system, so this method

175

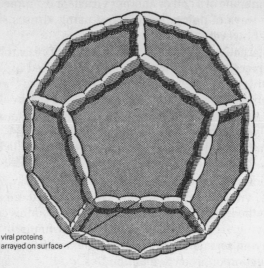

viral proteins
arrayed on surface

Figure 15. ISCOMS – the name is short for immunostimulatory complexes – have viral proteins arranged on their surface. ISCOMS are about the same size as viruses.

may make it possible to boost the neutralizing antibodies to HIV to a level where these provide effective immunity. Tests on experimental animals have shown that ISCOMs can improve the antibody response to a particular antigen a hundred-fold. The next question is what would happen if infected people received such a vaccine. If they too could produce much higher levels of neutralizing antibodies these might allow them to combat the infection.

In September 1987 Jarrett said that he hoped to be ready to start testing a vaccine against HIV based on ISCOMs within a year. Other British researchers, however, have warned against pinning too much hope on this approach. Geoffrey Schild, the director of the British programme of research on AIDS, has said: 'We should not underestimate the scientific complications of a vaccine against HIV. There is no cast-iron evidence that antibodies to the envelope are helpful. We need to look at the cellular immune response, and we still have to cope with virus variation.'

Indeed, some scientists believe that a vaccine based on the envelope protein gp120 may be positively harmful. There is some suspicion that gp120 may be toxic because part of it is similar to the nerve growth factor, neuroleukin. This resemblance may explain the effects of HIV on the brain, such as dementia (see p. 113). Other critics point to potential repercussions within the immune system. Some believe that the body does not produce effective antibodies against HIV because these would recognize human proteins. The result would be autoimmune disease, when the immune system starts to attack elements of the host's body, believing them to be foreign. Another theory is that a vaccine based on gp120 could generate antibodies that recognize the viral binding site – as well as a second generation of antibodies capable of binding to the receptor on the T-helper cell (see p. 179). These antibodies – which would resemble part of the viral protein and would bind to T-cells – might themselves cause immunosuppression.

Concern that antibodies to gp120 may be of little use in combating the infection has led other researchers to look more closely at antibodies to other elements of the virus. The level of antibodies to the core protein of HIV, called p24, for example, tends to fall at the time that the individual progresses to AIDS. Some scientists have suggested that if it were possible to maintain levels of these antibodies, this might prevent the development of AIDS in people who are infected with the virus but have not yet developed symptoms. This argument may be flawed. There have been several reports that, in African patients, there is no decrease in p24 antibodies before the individual progresses to AIDS. So p24 antibodies may not have a protective role after all.

Perhaps the biggest obstacle to a vaccine against HIV is the variability of the virus, particularly the outer envelope protein. Yet some components of this protein must stay the same in all strains of the virus. If the region where the virus binds to the T-helper cell were to mutate, the virus would no longer be able to attack its target. Studies of the protein gp120 have shown that there are a few sites where the sequence of amino acids remains the same in several strains of HIV, as well as in the virus which causes AIDS in monkeys, SIV. One or a group of these so-called 'conserved' regions must form the binding site.

In molecular terms, the binding site is quite large. One way to study it is to map the whole of the envelope protein to try to identify which parts are concerned with binding. A team of researchers at Cambridge University in England, led by Peter Lachmann, is following this line of inquiry. Their basic material is eighty-six different peptides (short segments) of the protein gp160. Each peptide is fifteen amino acids long, and overlaps with its neighbours on either side, so that every part of the protein is represented twice.

Lachmann and his collaborators began in 1987 to inject each peptide into rats. The rats produce antibodies against the foreign peptides. Having isolated these antibodies, the researchers planned to test these to find out which ones recognize the binding site on HIV. They will then know which parts of the protein are involved in binding to the T-helper cell. Theoretically, the antibodies that the animals make against these parts of the protein should be capable of preventing the virus from binding to the T-helper cell.

Lachmann told *New Scientist* in August 1987 that if any of the peptides prove to be a good candidate for a vaccine, he would then test them on himself and his colleagues to see how good they are at raising antibodies in humans. If the response is good, he would then find out whether these peptides were capable of preventing deliberate infection with the virus in other animals, possibly chimpanzees.

Other British and American researchers are also closely scrutinizing the binding site of the envelope protein. They are probing the potential of what some leading scientists believe is the best prospect for a vaccine: 'anti-idiotype anti-

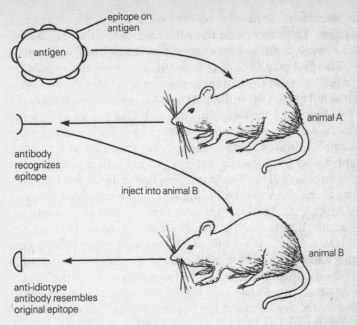

Figure 16. Anti-idiotype antibodies are antibodies against anti-bodies. If you inject an antigen into animal A, the animal makes antibodies against that antigen. If you then purify those antibodies and inject them into animal B, this animal recognizes them as foreign and makes its own antibodies against them. Some of these antibodies are anti-idiotype antibodies: they resemble the original antigen.

bodies'. These antibodies may also prove effective as a therapy for people who are already infected with the virus.

An anti-idiotype antibody is an antibody to an antibody (see figure 16). What makes each kind of antibody unique is a specialized region of the molecule that recognizes the shape of a particular antigen. This part of the antibody is called the 'idiotype'. It binds to the part of the antigen called the epitope. An antigen such as a protein has many epitopes on its surface.

If researchers take antibodies from one animal, and inject them into a second animal, the second animal recognizes the antibodies as foreign antigens. It makes its own antibodies against them. Some of these antibodies are anti-idiotype antibodies. They recognize the idiotypes of the first antibodies. The result is antibodies shaped like the original epitope.

The first results of applying anti-idiotypes to the problem of HIV were encouraging. Researchers in the US and Britain injected mice with antibodies against the CD4 receptor on the T-helper cell – the molecule to which HIV binds. They found that the mice produced anti-idiotype antibodies in response. Some of these scientists then discovered that anti-idiotype antibodies produced in this way can neutralize several different strains of HIV. The antibodies probably do this by attaching themselves to the binding site on the envelope protein.

Most importantly, from the point of view of developing a vaccine, they found that these anti-idiotypes could block both AIDS viruses, HIV-1 and HIV-2, as well as the related virus that causes AIDS in some monkeys, simian immunity virus. Work by researchers in London also suggests that the binding site on all three viruses is highly conserved. In other words, the molecules that form the binding site are the same in many different strains.

To obtain really precise information about what is going on, however, some of these researchers intend to study the viral binding site with monoclonal antibodies. Monoclonal antibodies are produced by genetically identical cells – clones – manufacturing quantities of a single type of antibody which can then be purified. For this work, researchers will use a series of monoclonal antibodies, each one against various epitopes of the CD4 receptor on the T-helper cell.

The next step in this complicated process is to inject each monoclonal antibody into mice, one at a time. By isolating cells that produce anti-idiotype antibodies against the first antibody, it will be possible to make monoclonal anti-idiotypes. Each of these will bear a resemblance to particular epitopes on CD4.

It will also be possible to manufacture a range of cells. Each culture will contain cells bearing a different peptide of the envelope protein gp120 on their surfaces. The researchers will then be able to add different monoclonal anti-idiotypes to each type of cell. By observing which antibodies bind to which cells, they anticipate being able to find out which portions of the envelope protein each antibody recognizes.

The aim of this kind of work is to map the viral binding site. Once scientists know the structure of the binding site, they may be able to design peptides or other molecules that would block it. They may well find that a single monoclonal anti-idiotype will not block the binding effectively. It might be that a pool of several monoclonal anti-idiotypes, each recognizing different sites, may be better at preventing the virus from binding to the T-cell.

This work may be vital in designing a vaccine. Yet it may also be possible to employ similar strategies as a therapy for people already infected with HIV. British researchers have been pursuing this line of inquiry. They first showed that a monoclonal antibody made against an epitope of CD4 called Leu-3a can effectively inhibit viral replication in the laboratory. The researchers aimed to give this preparation to patients with severe signs of HIV infection, to see whether it can also stop the virus replicating in the body.

The underlying theory is that the injected antibodies could block the CD4 binding site, so preventing the virus from attacking the T-helper cell. One theoretical objection to this strategy is that this type of blocking could itself cause immunosuppression. Normal activation of T-helper cells by antigens would not be able to take place. On the other hand, some commentators believe that it may be helpful to prevent stimulation of T-helper cells, as this can sometimes activate the virus. In any case, uninfected volunteers have already been injected with Leu-3a without ill effects.

The logical sequel to this work is to present the mouse antibodies in such a way that the body makes its own anti-idiotypes to them. These anti-idiotypes would recognize the viral binding site and, hopefully, block it. But the problem here is that the immune system only responds to antigens presented in a particular way. To provoke an immune response, the mouse antibodies would have to be 'flagged' with a substance called an adjuvant (see figure 17).

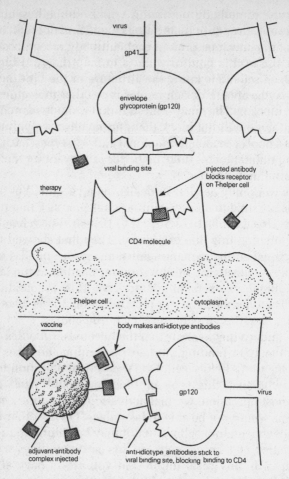

Figure 17. One potential strategy for the treatment of HIV infection may be to inject antibodies that recognize part of the receptor molecule on the T-helper cell. This would prevent the virus from binding to the cell. Alternatively, if such antibodies were injected as a vaccine, in a form that stimulated the body's immune system, the body might produce anti-idiotype antibodies that would block the envelope protein of the virus. These, too, would prevent the virus from attacking the T-helper cell.

Why do scientists think that they may be able to get the body to produce effective antibodies when its own immune system fails in this task? Luc Montagnier of the Pasteur Institute in Paris has suggested reasons why the portion of the virus that binds to the T-helper cell provokes such a meagre immune response. First, the sugar molecules that cover the envelope protein may mask its binding site. Alternatively, the binding site might be concealed in some other way on the surface of the virus. Finally, and most frighteningly, the binding site of HIV might mimic proteins present elsewhere in the body. The immune system may be fooled into thinking that these camouflaged molecules belong to the host, so it does not attack them.

There is some evidence that this possibility may be true. Some experiments suggest that the binding site of HIV resembles other proteins in the body. This would explain why the body fails to mount an immune response against the binding site.

Despite this potentially worrying possibility, the anti-idiotype approach is among the most promising developments. If such a vaccine worked, it would be inexpensive to manufacture, and it would not involve potentially toxic components such as another virus. Above all, if it is effective, it would produce antibodies that would neutralize many different strains of the virus.

The development of a vaccine is difficult enough; testing one that looks as though it has potential raises yet more complications. Traditionally, scientists have tested new vaccines on large populations of people at risk of catching the disease in question. They then establish whether the group that had the vaccine suffered fewer bouts of disease.

This approach is possible with a disease such as whooping-cough or smallpox. In the absence of a vaccine against these diseases, there was little that people could do to avoid them. AIDS is different. People taking part in a controlled trial of a vaccine to protect them against HIV infection would have to be fully informed. They would need to know about the potential risks of a novel vaccine. Just as importantly, there would be an ethical obligation on the part of the researchers to educate the participants about how to avoid HIV infection. This requirement would immediately make the trial a less powerful means of determining the effectiveness of a vaccine.

Another possibility would be to test a vaccine in known high-risk groups, such as intravenous drug users, prostitutes or homosexuals, depending on the pattern of infection in the country concerned. Again, there would be a responsibility to educate these groups about reducing their risks. Drug users are notoriously difficult to influence via conventional educational programmes. This characteristic might make them appear to be suitable people to test a vaccine on, if investigators could cope with the ethical dilemma involved. But it would also be difficult to trace and to follow up drug users, not to mention the practical difficulties of enrolling people in a trial because they participate in what is, in many countries, an illegal activity.

Another problem will be how to distinguish between people who are infected and those who have produced antibodies in response to a vaccine. Vaccinated people might find it difficult to get life insurance, for example, if they had no means of proving that they are not infected. One proposal to avoid this difficulty is to include some unusual antigen, to which people would not normally be exposed, with the vaccine.

Initial tests of candidate vaccines have already begun. But the scientific community has yet to agree on criteria that will help to ensure that tests are fair and safe. In Britain, leading authorities in AIDS research have called for a set of international guidelines for vaccine tests. Without such guidelines, the evaluation of potential vaccines may go down the same road as that of drugs for AIDS. Such mismanagement, were it to happen, could add years to the time it takes to establish the safety of a vaccine. A great deal of resources, and, most importantly, many lives, would be lost in the process.

CHAPTER 12

THE AFRICAN DIMENSION

When AIDS began to appear in Kinshasa, the capital of Zaïre, the first people to suffer were prostitutes in their late teens and early twenties. A little later, it was the turn of their clients, typically men in their early fifties – the 'sugar daddies', as they are commonly described. Then, babies born to infected mothers began to fall ill. Eventually, doctors became aware of AIDS in young children, infected by contaminated blood in transfusions.

The story of AIDS in Kinshasa illustrates the differences between how AIDS has spread in Africa and how it has developed in the US and Europe. In much of Africa, homosexual activity and drug abuse are uncommon. In Africa, unlike the US and Europe, the human immunodeficiency virus is spreading mainly by heterosexual intercourse, by blood transfusions, and from infected mothers to their children.

The first hints that AIDS could be present in Africa as well as the US and Europe followed the observation in 1983 that a high proportion of the cases of AIDS in Europe involved people from Africa. Doctors from Brussels reported in the *Lancet* that they had identified a group of five black Africans with AIDS. These patients, from Zaïre and Chad, had been living in Belgium for between eight months and three years. Soon after, French researchers in Paris announced twenty-

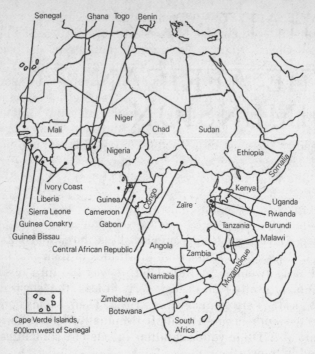

Figure 18. *Most of the African countries mentioned in this chapter are shown on this map. HIV-1 is most common in central and east Africa. In west Africa, HIV-2 is more common than HIV-1.*

nine cases of AIDS in France. These doctors found that most of their patients had travelled to the US, Haiti or equatorial Africa. Those who had gone to the US were mainly homosexual men; the rest were mainly heterosexual. The association with Africa was 'striking', the researchers said.

Another French team found AIDS in a twenty-three-year-old black woman who had left Zaïre in June 1981. She died of her infections in March 1982. The scientists warned that, along with evidence of AIDS in Haitians and haemophiliacs in the US, this case suggested that the syndrome was not restricted to homosexuals and drug abusers.

Further evidence of an 'African link' began to emerge. A Danish woman, who had lived in Zaïre, had probably died of AIDS in 1977. This woman, a surgeon, worked at a rural hospital in northern Zaïre between 1972 and 1975. She then visited Ghana, Nigeria, Senegal and the Ivory Coast before resuming work in Kinshasa, the capital of Zaïre, where she stayed from 1975 to 1977. She had suffered repeated episodes of diarrhoea during her time in Africa. She had also lost weight and developed enlarged lymph nodes, a typical sign in HIV infection. In 1977 she had attacks of pneumonia caused by *Pneumocystis carinii*, from which she eventually died. The researchers who reported her case wrote in the *Lancet*: 'She could recall coming across at least one case of Kaposi's sarcoma while working in northern Zaïre, and while working as a surgeon under primitive conditions she must have been heavily exposed to blood and excretions of African patients. She had not been to the US or to Haiti, and did not abuse drugs.'

At the same time, doctors working in Belgium said that over the past couple of years, they had seen 'at least a dozen' patients from Zaïre with symptoms that suggested AIDS. The most prominent indication of AIDS in these patients was a fungal condition called cryptococcosis. Only wealthier people in Zaïre could afford to come for treatment in Europe, these doctors pointed out, so the cases they had seen were probably the tip of the iceberg.

These doctors also described the case of a thirty-four-year-old black woman from Zaïre, who, they said, had probably had AIDS in 1977. This woman, a secretary with an airline company, had had three healthy children by her first husband. She had remarried and had three more children by her second husband. Of these three, the two eldest had already died by August 1977 when she brought the third child from her second marriage to Belgium for medical treatment. While she was in Belgium, the woman began to suffer symptoms and infections typical of AIDS. Eventually, she flew back to Kinshasa, where she died in February 1978.

Later in 1983, evidence emerged that AIDS in Africa was not confined to Zaïre and Chad. French doctors described of the case of a man from Mali who had never been to central

Africa. In fact, he had lived and travelled only in Mali and France. His last visit to Mali was in 1982. He had had 'no special contact with anyone from Zaïre at home or at work'. This man had been married twice and he had four children. He did not inject drugs, he was not a haemophiliac, he had never had a blood transfusion, and, the doctors said, he was 'strictly heterosexual'. He first showed signs of illness in May 1981, eventually dying of infections typical of AIDS in June 1983. Mali is well over a thousand miles from Zaïre. AIDS, it seemed, was more widely distributed in Africa than people had first thought.

Nevertheless, the link with Zaïre grew stronger. In February 1984, Nathan Clumeck and his colleagues from the Saint-Pierre University Hospital in Brussels described AIDS in twenty-two black Africans and one Greek man who had lived in Zaïre. These patients all had either opportunistic infections or Kaposi's sarcoma, or symptoms such as enlarged lymph glands, fever, loss of weight and diarrhoea. In addition, they all had immunological abnormalities typical of AIDS. None was homosexual, none used intravenous drugs and none had had a blood transfusion. Sixteen of the eighteen patients with AIDS came from Zaïre; of the remaining two, one came from Burundi, which shares a border with eastern Zaïre, and the other was from Chad in the north. By looking at the medical records of these patients, the doctors found that the earliest date of diagnosis was May 1979, in a patient from Zaïre.

Clumeck and his colleagues suggested that, because the syndrome was appearing in young to middle-aged men and women, it might be spreading by heterosexual contact. They added: 'We believe that AIDS is a new disease that is spreading in central Africa.'

Whether AIDS was a new disease or one that had been left undiagnosed for many years was difficult to determine. Perhaps, some Western researchers suggested, AIDS had always been present in Africa. Some investigations, however, supported the idea that something new was happening. Apart from cryptococcosis, other diseases seemed to be behaving oddly. The skin tumour Kaposi's sarcoma, for example, had changed from being a fairly mild condition into a fatal one.

Up to the end of the 1970s, those people who suffered from Kaposi's sarcoma in Africa were mostly older men. They developed swelling and purplish nodules on the legs and feet. Often these tumours would go away on their own but, if not, powerful drugs would do the trick. Patients often survived more than ten years with this condition.

By 1983, this description no longer held good. Anne Bayley, professor of surgery at the University Teaching Hospital in Lusaka, Zambia, said in 1983 that Kaposi's sarcoma had changed. She said, 'Whatever I do, these patients all die.'

As Kaposi's sarcoma was associated with AIDS in American homosexual men, the next step was to find out if AIDS had any connection with this new type of Kaposi's sarcoma in Africa. Bayley examined and tested patients with Kaposi's sarcoma to see whether they had antibodies to the AIDS virus. Virtually all the patients with the unusual type of Kaposi's sarcoma had antibodies to the virus. People without Kaposi's sarcoma, or with the traditional variety, were far less likely to have antibodies to the virus.

In June 1984, Bayley wrote in the *Lancet* that she could now divide her patients with Kaposi's sarcoma into two distinct groups: those with the typical disease which responded to treatment, and those with unusual symptoms who died rapidly. From 1975 to 1982, Bayley used to see up to about a dozen new patients with Kaposi's sarcoma a year, most of them with typical symptoms. By 1983, the pattern had changed. In that year, ten new patients came under Bayley's care, all of them conforming to the traditional pattern. They all got better when treated with drugs. But there was a second group of thirteen patients who appeared that year. These individuals were younger than the other group and better educated. Four of them, Bayley said, had similar jobs, suggesting an occupational or social cluster. The key characteristic, however, was that their Kaposi's sarcoma progressed rapidly and did not respond well to treatment. It was, in other words, very similar to the type of tumour suffered by immunosuppressed homosexuals.

At around this time, doctors in Britain described the case of a woman from Uganda who had developed Kaposi's sarcoma. They wrote in the *Lancet*: 'To our knowledge, this is the first case of AIDS in a patient from Uganda.'

During 1983 and 1984, teams of researchers began studies to determine how widespread AIDS was in some African countries. One group, led by Peter Piot of the Institute of Tropical Medicine in Antwerp, in Belgium, examined patients at hospitals in Kinshasa, Zaïre. During three weeks in October 1983, they identified thirty-eight patients with AIDS.

Of these thirty-eight, the men tended to be older than the women. Men and women were affected almost equally: the male to female ratio was 1.1:1, suggesting that the disease was spread heterosexually. There was also some evidence that relatively more cases might be occurring in people with higher incomes.

Out of fifteen men questioned, thirteen said that they had had sexual intercourse with more than one woman during the year before their illness started. These men had, on average, seven sexual partners a year, though some had had as many as a hundred. Eight women patients also provided information on their sexual partners. Six of them had had more than one partner during the year before the onset of symptoms, with an average of three and a maximum of five.

Piot and his colleagues estimated that there were about 17 cases of AIDS every year for every 100,000 people living in Kinshasa. (For comparison, about 14 in every 100,000 single men in the US had AIDS in 1984, rising to about 340 per 100,000 in San Francisco.) If children were excluded from the population, the rate would be even higher.

These researchers also wondered about the possible origin of the disease. Although exact figures were not available, they knew that several thousand professional people had gone from Haiti to Zaïre between the early 1960s and the mid-1970s. After talking to Haitians still living in Zaïre, the scientists concluded that most of those people had now left Zaïre and gone to live in Europe or the US. 'We were unable to identify any common factor accounting for the occurrence of AIDS in Haiti and in Zaïre almost simultaneously,' they said. They knew of only one case of AIDS among Haitians living in Zaïre – an unmarried woman in 1983. So, they concluded, they knew of no facts implicating either central Africa or Haitians as the origin of the disease.

Piot did not know when AIDS first appeared in Kinshasa. As far back as 1975, there seemed to have been some cases of weight loss, swollen lymph glands and aggressive Kaposi's sarcoma in young adults, but there was not enough information to establish these cases as AIDS. Despite the reports of probable cases of AIDS in Zaïre in 1976 and 1977, it was unlikely that African patients with AIDS seeking treatment in Europe would have gone unrecognized.

This group of researchers also noted a sharp increase in the number of cases of the fungal infection cryptococcosis in Kinshasa. Cryptococcosis is a common infection in patients with AIDS, especially those from Africa. Between the mid-1950s and late 1979, each of two hospitals in Kinshasa had, on average, one case of this infection a year. Between 1981 and early 1984, the two hospitals between them had seen more than thirty-five cases of cryptococcosis, most of them thought to have been probably associated with AIDS. Piot and his colleagues concluded: 'This is consistent with the emergence of the disease in large numbers simultaneously with the first cases in the USA and Haiti.'

While research continued in Zaïre, other scientists studied the situation in Kigali, the capital of Rwanda. Philippe Van de Perre and his colleagues from the Saint-Pierre University Hospital in Brussels diagnosed twenty-six patients with AIDS during a period of four weeks in late 1983. They estimated that there were probably eighty cases of AIDS a year for every 100,000 people living in Kigali.

Two of the twenty-six patients diagnosed were children, both aged under two years. Of the twenty-four adults, seventeen were men and seven were women. Thirteen of the men said that they had had frequent and regular heterosexual contacts with different partners, including prostitutes in eleven cases. Three of the seven women were prostitutes; two others said that their husbands frequently went to prostitutes. So, as the researchers said, 'sexual promiscuity could be a risk factor among heterosexual patients with AIDS'. Many of the patients, in fact, had evidence of other sexually transmitted diseases such as gonorrhoea and syphilis.

Factors such as homosexuality, use of intravenous drugs and blood transfusions did not seem to figure in these cases. However, Van de Perre found that almost all the patients were middle or upper class. They concluded: 'Urban activity, a reasonable standard of living, heterosexual promiscuity and contacts with prostitutes could be risk factors for African AIDS.'

In December 1984, a local paper in Uganda published an article on an unusual illness. People in the Rakai district of Uganda, which borders the western shores of Lake Victoria, near the border with Tanzania, were wasting away and dying of diarrhoea. The local name for it was 'slim disease'. Doctors in Uganda wondered if the new illness was related to the syndrome of weight loss and diarrhoea that their colleagues in the US were reporting in homosexuals with AIDS.

The government sent a team of investigators to find out what was happening. They said that the illness did not seem to be due to outbreaks of cholera or typhoid. Then, in June, doctors in Uganda decided to mount a weekend expedition to Masaka, on the edge of the Rakai district. One of them was Anne Bayley, who was in Kampala as an external examiner for the finals exams at the medical school.

The group eventually arrived at a hospital in Masaka. Of 110 patients on the wards, the doctors could diagnose twenty-nine as having AIDS from their symptoms alone. The doctors also took blood samples. Subsequent testing showed that all these patients were positive for antibodies to the AIDS virus.

The results of this study, together with those of a survey of forty-two patients with slim disease in Kampala, most of whom also came from Masaka or the Rakai district, appeared in the *Lancet* in October 1985. When the researchers tested blood samples from the patients in Kampala who had slim disease, they found that about three quarters of them had antibodies to the AIDS virus, too. They also tested samples of blood from 410 healthy medical staff at Mulago Hospital in Kampala; 10 per cent of them were positive.

Although the doctors described slim disease as a new syndrome, previously unreported in Uganda, it later became clear that this condition was indeed AIDS. Slim disease seemed to be a new disease in Uganda. Post mortem medical records in Uganda are good, going back to 1944. The researchers said that if slim disease had been present earlier, they would have been able to find reports of it in the records.

The first recognized cases of slim disease in Uganda occurred in a small fishing village on Lake Victoria, just north of the Tanzanian border. Villages such as this one became host to traders and smugglers in the days when road blocks in Uganda prevented goods from flowing out of the country into Tanzania. Some of the first people to suffer from slim disease were the traders. The traders may have brought the virus back with them from Tanzania.

Bayley and her colleagues speculated about how the virus could have entered Uganda. 'The notion that the disease may have been transmitted sexually from Tanzania is interesting since it fits historically with the movements of the Tanzanian army in 1980 and the subsequent movements of the Tanzanian traders. Of the fifteen traders tested for evidence of . . . antibodies, ten were positive. These traders admitted to both heterosexual and homosexual casual contacts. Tanzanian soldiers entering Uganda since 1980 have had frequent heterosexual contact with the local population.' But the puzzle remained. 'If the virus did come from Tanzania, where did Tanzania get it from?'

By 1985, many new tests for antibodies to HIV had became available to researchers who wanted to study the extent of infection in Africa. Teams of scientists mounted expeditions to remote areas to find out how far the virus had spread. Others carried out tests on samples of blood serum that had been taken years before and stored frozen. They wanted to find out exactly when the virus had appeared in Africa. The figures that began to appear at first sight suggested, rather surprisingly, that the epidemic was far older and more extensive than anyone had first thought.

For example, Carl Saxinger of the National Cancer Institute in the US, and his colleagues, found, in a study that would later be discredited, that over 60 per cent of healthy children in the West Nile district of Uganda had, in the early 1970s, had antibodies to the AIDS virus. Saxinger and his co-workers wrote in Science, '. . . it is likely that residents of the West Nile region of Uganda have been and continue to be exposed to the virus at a very early age'. They also suggested: 'It is possible that AIDS existed in African populations without being recognized as a separate disease entity. The virus may have originated in Africa in the past and exposure to the virus may be much more common than AIDS itself in some populations.'

Robert Biggar, also of the National Cancer Institute, published the results of several studies carried out around 1984. In a remote part of eastern Zaïre, for example, he found that over 12 per cent of 250 Zaïrean hospital outpatients had 'clearly positive' results to tests for antibodies to the virus. According to Biggar, another 12 per cent had 'borderline' results. Yet none of these patients had symptoms suggesting AIDS. Furthermore, no case of AIDS had ever been diagnosed at the hospital where these tests were carried out.

Biggar, together with American and Kenyan collaborators, also tested samples of blood taken from people all over Kenya during 1980 and 1984. Overall, about one in five of almost seven hundred people tested seemed to have antibodies to the virus, with another 24 per cent of those tested having 'borderline' reactions. Antibodies to the virus seemed to be more common in people from some parts of Kenya rather than others. For example, half of the ninety-nine people tested from the remote Turkana district had positive results.

There were warning signs that something might be wrong with these results. In the Kenyan study, the researchers mentioned that they knew of only three cases of AIDS in Kenya, all diagnosed in Nairobi. As the quality of medical care in Kenya was quite good, it seemed odd that cases of AIDS were not appearing at a rate which would reflect the number of people in the population apparently affected.

Biggar and his colleagues speculated on why this should be the case. Perhaps the manifestations of AIDS in Africa were different to those in the US and Europe. Perhaps the course of infection was different in Africa. If people became infected as children, maybe the virus did not kill them; if it did, perhaps the people with antibodies were those who had survived infection. There was a third theory: perhaps infection alone did not cause abnormalities of the immune system. Some other factor which caused AIDS to develop in infected people in the US and Europe may have been missing in Africa. Finally, Biggar suggested, maybe the virus was just less virulent in Africa.

In fact, the cause of these odd results was probably that blood serum from Africans reacted with the proteins present in the tests, even though these people had no antibodies to the AIDS virus. In other words, they were false positives.

Biggar revised his opinions. In early 1986, he wrote that knowledge of the spread of AIDS in Africa had been hampered by 'lack of data and perhaps misleading preliminary laboratory results'. When he tested for the second time forty-six healthy people from Zaïre and Kenya whom he had originally believed to be infected with the virus, he found only two that had results typical of true positives. Surveys in Africa, he said, had probably greatly overestimated the extent of the spread of HIV. Similarly, tests on repeatedly frozen and thawed blood samples taken from African children more than a decade ago – as in Saxinger's study – had probably also resulted in overestimates.

Biggar reviewed the evidence that AIDS was, in fact, a new disease in Africa. 'In my conversations with clinicians practising in tropical Africa during the 1960s and 1970s, they have stated strongly that if AIDS had existed as anything other than rare, sporadic cases, it would have been recognized . . .' Similarly, cases of Africans seeking medical advice in Europe for diseases resembling AIDS only began to appear since 1980. Furthermore, the fact that AIDS is spreading in Africa 'strengthens the probability that it is a new disease'. And, Biggar added, 'There is no conclusive evidence that the AIDS virus originated in Africa, since the epidemic seemed to start at approximately the same time as in America and Europe.'

Such disclaimers helped to salve the wounds that some African countries felt had been inflicted by accusations that Africa was to blame for AIDS. Many Africans had also been taken aback by the sweeping generalizations about sexual behaviour 'in Africa' that were being bandied about.

There was much talk of 'promiscuity'. Rumours abounded, with little evidence to support them, that anal intercourse was common between men and women in Africa, either to preserve virginity or as a method of contraception. Such theories probably arose from a reluctance to admit that the virus spreads by conventional heterosexual intercourse.

Sexual behavior does, of course, greatly influence the spread of AIDS. In any society, the more sexual partners an individual has, the greater the likelihood he or she has of encountering someone with a sexually transmitted disease. One researcher succinctly summed up the range of human sexual behavior. First, he said, there are the 'inactives'. Then, there are those who are more or less monogamous. Next come the 'fast trackers', followed finally by the professionals. These categories probably apply to all societies; what differs, according to time, place and history, is the proportion of people in each category. Societies that have undergone great economic and political upheaval are likely to have different proportions of fast trackers and professionals than societies that have had a relatively stable recent history.

Social and economic factors have probably played a important role in the spread of AIDS in Africa. Many of the countries of Africa have suffered political upheaval and wars in the past few decades. Large movements of people – armies, refugees or those seeking jobs, for example – have probably played their part in widening the epidemic. In other places, it may have taken just a few highly sexually active people to sow the virus in a previously unaffected population. Geographic factors, such as lakes, mountains and rift valleys, have undoubtedly stood in the way of the virus. And, as one doctor said: 'Sexual isolation may have a lot to do with the quality of the roads.'

The virus has certainly been spreading along trade routes within Africa. The port of Mombasa is the point of entry for many goods destined for Kenya, Uganda, south Sudan, north-east Zaïre, Rwanda, Burundi and part of Tanzania. A main road leads from Mombasa, passing through Nairobi, around the top of Lake Victoria, and down through south-western Uganda (see figure 19). It skirts the northern border of Rwanda and carries on into Zaïre, eventually ending up at Kinshasa on the Congo river. Between Kampala and the Ugandan border with Zaïre, the road passes through a small town called Lyantonde.

Lyantonde is one of several small towns along the road that exist to cater for the needs of the truck drivers who transport goods to and from the ports. With a population of about four or five thousand people, it has about thirty bars and lodges. At dusk, lorries and trailers line the main street. In Lyantonde, the lorry drivers and their mates, called 'turn boys', can find a meal, beer and women.

Many of the women work as 'bar girls'. They are essentially prostitutes working in bars, where they get food and a uniform. Most of them are aged between about sixteen and twenty-five and come from the surrounding countryside.

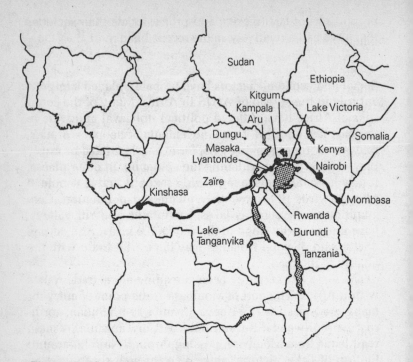

Figure 19. The road that runs between Mombasa on the east coast and Kinshasa in the west. Some of the towns and villages in Zaïre and Uganda that are mentioned in this chapter are also shown.

They usually have at least one sexual partner a day, receiving in return, if not money, goods such as paraffin or sugar.

Between 1986 and 1987, researchers in Uganda tested almost two hundred of the bar girls for antibodies to HIV. About two thirds of them were positive. For comparison, the researchers also tested mothers and their children attending an immunization clinic. About 17 per cent of these women were infected, as well as just over 1 per cent of the children.

The next group to have blood tests comprised seventy-four lorry drivers and turn boys. About a third of them were infected with the virus, too. Most of these men had been to most of the countries served by the port of Mombasa, and most had had sexual partners in each country. The majority of these individuals had had more than fifty sexual partners in their lifetime; 80 per cent of the rest had had more than ten. The researchers also went back later and tested the negative men with a direct test for the virus itself, rather than for antibodies. When they took these results into consideration, up to half of the drivers and their mates had signs of exposure to the virus.

Zaïrean and French researchers have also shown that individuals more likely to have contact with outsiders have higher rates of infection. Aru is a village on a busy road in Zaïre, on the border with northern Uganda. It has a population of about 10,000. The researchers analysed tests on over three hundred people according to where they lived and whether they came from outside the area or from the established ethnic group in the town. They found that some people from the busy market district of the town had antibodies to the virus, compared with none from the less crowded areas. Very few of the long-term residents in Aru were infected, compared to over 10 per cent of people who had travelled in from other parts. One interpretation of these results could be that people drawn to Aru by the opportunities for trade and profit might have brought the virus with them. HIV, at the time of this study in 1986-7, was beginning its insidious spread into the people native to the area.

Other studies have shown that people from the cities are spreading the virus into the countryside. In January 1986, Belgian researchers carried out a survey in Dungu in northeastern Zaïre. This area, said the researchers, is 'a remote and sparsely populated region with no documented AIDS cases. The Azande [ethnic group] living there are a declining population with high female infertility ascribed to venereal disease.' In Dungu, the researchers tested over two hundred adults not suspected of having disease due to HIV. About a quarter of the people who had 'strong links' with the capital of the area, Isiro, had antibodies to the virus. In comparison, fewer than one in twenty of those who rarely travelled from Dungu had positive results.

The researchers also went to a nearby village, Ndedu,

which was even more isolated from the capital of the area. Only one of 222 adults and none of 170 children there had antibodies to the virus.

The scientists concluded: 'This is a clear model of urban-to-rural spread of HIV in a remote area previously unsuspected of HIV infection.'

With AIDS in Africa, however, it is difficult to generalize. By 1986, in some countries, the virus was already beginning to spread in rural areas with little contact with the towns. In Burundi, researchers carried out a study in an area of about 120 square kilometres, called Butezi, which is inhabited mainly by peasants. The roads in Butezi are nearly non-existent, and even communications with the capital of the district, Ruyigi, are difficult. The researchers collected blood serum in 1986 from 158 outpatients attending the Butezi hospital for testing. These scientists found that nearly twenty per cent of sixty-five people who had sexually transmitted diseases (STDs) had antibodies to the virus. However, only four out of ninety-three patients attending the hospital for treatment of other complaints, such as malaria, hepatitis, tuberculosis and urinary problems, were positive – a rate of just over 4 per cent. The researchers concluded: 'Our results confirm a higher prevalence of HIV infection in subjects with a history of STD and indicate a rapid spreading of the virus even in a rural area with little or no contact with the urban areas.'

AIDS was by 1986 also affecting rural areas of Uganda. Kitgum is in a remote part of Uganda, about 200 miles north of Kampala. Researchers tested blood samples taken in 1984 and 1986. Out of 111 samples taken in 1984, only one was confirmed as positive. Tests on 491 samples taken in 1986 showed a different picture, however. Over 13 per cent of these people were positive. Summarizing their results, the researchers said: 'The great increase of anti-HIV seropositivity in such a short time indicates the dramatic spread of the epidemic even in remote rural areas of central Africa.'

One of the problems that scientists face in trying to work out the overall picture of AIDS and HIV infection in Africa is that information is very patchy. This unevenness sometimes reflects the fact that little research has been done; or it may indicate that governments are refusing to release the results, fearful of bad publicity that may, for example, affect their tourist trade. Whatever the reason, it is not possible to generalize from country to country. In 1986, for example, a survey showed that the incidence of infection in a remote part of Kenya was very low – only about one in a hundred people there were positive.

The search for the virus throughout Africa continued. By 1985, there had been only a handful of reports of AIDS in west Africa. An odd feature of some of these cases was that, although these people had symptoms typical of AIDS, when doctors tested their blood for antibodies against HIV, the results were negative. Scientists from Lisbon, in Portugal, and from Paris, including Luc Montagnier, decided to take a closer look at two such patients from West Africa.

One of these individuals was a man from Guinea-Bissau, who began to suffer symptoms of diarrhoea and loss of weight, as well as swollen lymph glands, in 1983. He also had infections typical of AIDS. The second patient was born in and lived in the Cape Verde islands, 500 kilometres off the coast of Senegal. He had first developed symptoms in January 1982; doctors in France diagnosed AIDS in June 1983. Tests for antibodies to HIV in both these patients were repeatedly negative.

The scientists managed to isolate viruses from both men. Powerful photographs of infected cells, taken using an electron microscope, showed a virus with an unusual appearance: instead of the viral membrane appearing relatively smooth, it was covered with spikes. The virus was unlike HIV, and was later called HIV-2.

Later, many of the same researchers reported another thirty cases of AIDS caused by HIV-2, almost all of them in people from west Africa. Many of these people came from Guinea-Bissau, and two came from Cape Verde. One was an eleven-year-old boy from Angola who had lived in Cape Verde for several years. Finally, there was a forty-year-old Portuguese man who had lived for eight years in Zaïre and who said he had never stayed in west Africa. From the characteristics of these patients, it seemed that HIV-2, like HIV-1, was transmitted mainly by heterosexual contact.

A new epidemic of AIDS could be about to occur in west Africa, the scientists warned, this time caused by HIV-2. This virus was spreading rapidly in west Africa. A survey of 275 people working at the Ministry of Health in Guinea-Bissau took place in 1986 and 1987. Almost one in five of the women were positive for HIV-2, but fewer than one in fifty had antibodies to HIV-1. Among the men, over one in ten had antibodies to HIV-2 but none had signs of infection with HIV-1. In fifty-four healthy blood donors in Guinea-Bissau, only a couple were positive for HIV-1, while more than one in four had antibodies to HIV-2. The only known risk factor for those who were infected with either virus was contact with female prostitutes.

Both HIV-1 and HIV-2 seemed to have been present in the population in Guinea-Bissau for some time. The researchers tested 300 samples of blood taken and stored in 1980. In three (1 per cent) they found antibodies to HIV-1; a further six (2 per cent) were positive for HIV-2.

Surveys in other west African countries, such as Guinea Conakry, Ivory Coast and Benin have generally shown low levels of infection with HIV-1 and HIV-2 of the order of 1 or 2 per cent, and sometimes less, in the general population. Yet, as in central Africa, prostitutes seem to be a group at high risk of infection with HIV. In Ivory Coast, between 16 per cent and 65 per cent of prostitutes tested had antibodies to either HIV-1 or HIV-2.

In central Africa, the ratio of infected men to infected women is usually close to 1:1. In Ghana in 1986, however, the ratio was almost 12:1. By 1987, the ratio had dropped to just over 6:1, which suggested to researchers that the infection was still spreading among susceptible people.

Many researchers believe that AIDS caused by HIV-2 has

a much longer incubation period than that caused by HIV-1. For example, there is the case of a Portuguese man and his wife who were diagnosed as having AIDS due to HIV-2 infection. The man had done his military service in Guinea-Bissau between 1966 and 1969. Doctors investigating the couple suggested that AIDS may have taken between sixteen and nineteen years to develop in the man and eleven years in his wife. Another case, although the evidence for a long incubation period is less clear, occurred in a Portuguese man who died of AIDS in 1980. Between 1968 and 1974 he had lived and worked in Angola, both in the navy and as a truck driver from Angola to Mozambique. He first developed signs typical of HIV infection in February 1977, and went on to suffer infections typical of AIDS. Scientists tested a sample of his blood which they had stored in 1979, for antibodies to both HIV-1 and HIV-2. Antibodies to the envelope of HIV-2 were present.

Researchers who looked at the pattern of the first recognized cases of HIV-2 infection were struck by a common characteristic. Many of the infected people were either Portuguese, or from African countries linked by the former Portuguese trading routes. By 1987, cases of infection with HIV-2 had been reported in several African countries, including Cape Verde, Ghana, Guinea-Bissau, Guinea Conakry, Ivory Coast, Mali, Niger, Mozambique and Senegal. Mali and Niger, although not on the coast, are connected to it by the Niger river. Some cases have also had connections with Angola, another former Portuguese colony.

Some scientists believe that this pattern may provide a clue to the origin and subsequent spread of HIV-2. In previous centuries, the Portuguese ruled the seas. Cape Verde, for example, was an obligatory stop for ships going to South America. Researchers have suggested looking for HIV-2 in other places that the Portuguese visited, such as Goa (in India), Brazil and Cuba. By 1987, there had already been one report that HIV-2 was present in Brazil.

Conjecture about where the AIDS virus came from – there was only one virus in those days – has abounded ever since doctors first identified the existence of the epidemic. When researchers discovered a virus that caused AIDS in macaques in captivity, and then found that wild African green monkeys seemed to have antibodies to the same virus, although they did not develop AIDS, some people believed that the mystery of the origin was solved. The virus must have passed from monkeys to humans somewhere in Africa. It probably infected people living in some remote part of the continent. The disease might have remained undiscovered until social or political changes brought individuals from these isolated communities into contact with the towns. Alternatively, these people may have evolved some immunity to the virus. Only when the virus encountered people without any resistance to it did it begin to spread with a vengeance.

Such a theory admirably suited the purposes of those who wanted a neat and tidy answer to the problem. Yet, by the late 1980s, many researchers are agreed that the evidence for AIDS originating in Africa is weak. There are two lines of argument. According to the first, the fact that the viruses isolated in Africa are more variable than viruses isolated elsewhere could suggest that HIV has been present in Africa for longer than it has in the West, because the virus is likely to mutate more rapidly the more it passes from person to person. On the other hand, this variability could merely reflect the genetic heterogeneity of the people living in Africa, to which the virus has had to adapt.

The other type of evidence comes from 'retrospective' studies that involved studying past medical records to see whether earlier cases of AIDS had gone unrecorded. Most researchers agree that AIDS appeared almost simultaneously in the US, Haiti, Africa and Europe, during 1980–81. Retrospective studies did, however, identify some individuals who appeared to have died of a syndrome similar to AIDS before that period. In the US, most of these cases dated back to about 1977. In Haiti, the year was 1978. In Zaïre, in central Africa, there seemed to have been cases of probable AIDS in 1976 and 1977.

Several reports also appeared soon after the first descriptions of AIDS in the US, documenting similar cases earlier in Europe. One was a Spanish homosexual man who fell ill in mid-1981. Although he had a regular sexual partner, he had had sexual intercourse with other men in New York in 1974 and in Turkey in 1980. Another case that sounded like AIDS had occurred in a male homosexual violinist who had travelled widely in Europe, where he had many sexual contacts. As far as his doctors knew, he had not travelled to the US, Haiti or Africa. He fell ill in December 1976, and died of his disease in early 1979. And of the first twenty-nine cases of AIDS in France, doctors had seen nine before June 1981, when the first report of the syndrome appeared in the US. The earliest case was a thirty-one-year-old French homosexual man who had not travelled abroad in the five years before diagnosis. Doctors had diagnosed his Kaposi's sarcoma in 1974. Other reports of likely cases of AIDS came from Denmark in 1981.

Perhaps one of the most puzzling early cases – and one that doctors have confirmed with laboratory tests – is that of a sixteen-year-old boy who died of AIDS in 1969 in St Louis, Missouri, in the US. Known as Robert R., this teenager developed swollen lymph glands, lost weight and suffered severe infections. He also had Kaposi's sarcoma. Doctors were so curious about his death that they froze samples of his blood and tissue. Tests in 1987 showed that he was infected with HIV. Perhaps, researchers speculated, the virus had arrived in the US several times before, or been there all along. But it had never taken hold before in the way that it did in 1981.

In Africa, doctors are sure that if cases of AIDS had appeared in any numbers, they would have noticed them. Doctors in Uganda reported in 1985 that there had been a few cases of the atypical form of Kaposi's sarcoma as far back as 1962, 'and this could suggest that AIDS has been present since then'. Yet studies such as that by Saxinger, which suggested that in the early 1970s over 60 per cent of children in the West Nile district of Uganda had antibodies to HIV, have been proved wrong. Tests in 1986 on elderly people in Kampala, and on healthy adults from the West Nile district of Uganda, showed that none and less than 2 per cent respectively had antibodies to HIV. These results suggested that the disease had arrived in Uganda only recently.

Other researchers tested blood samples from over 6,000 people from nine African countries, including Zaïre, Uganda and Kenya, which had been taken and stored between 1976 and 1984. Only four samples contained antibodies to HIV. These scientists said: 'Our data do not support (nor do they totally disprove) an African origin for the human immunodeficiency virus. They do show that the virus has not been endemic in rural areas of sub-Saharan Africa until recently.' Another group of scientists tested over 1,200 blood samples which had been taken from people in central Africa years before, about 800 of them taken as long ago as 1959. The researchers found antibodies suggestive of HIV infection in just one sample taken from someone in Léopoldville (now Kinshasa) in 1959. Even if the virus was present at that time, it certainly was not at all common.

The idea that AIDS began in Africa is hard to abandon. The reason is that HIV-2 is closely related to the virus that causes AIDS in monkeys, SIV. And HIV-1 is related to HIV-2. One intriguing question, unanswered by 1987, is whether there exists a monkey virus as closely related to HIV-1 as SIV is to HIV-2. Could both HIV-1 and HIV-2 have crossed the 'species barrier' independently of each other? If so, why?

Perhaps the opportunity for this jump to happen arose as a result of closer contact between monkeys and humans arising from the demand for animals for scientific procedures. Some researchers have pointed out that there was a massive trade in monkeys from the 1950s onwards, from Africa, mainly to the US. The African green monkey is a species commonly used for some types of tissue culture. The capture, transport and care of these monkeys must have greatly increased the opportunity for contact with humans.

If HIV had existed in some remote tribal population, many

researchers believe, we would have found that population by now. Pygmies were the prime contenders for this theory. They live in the evergreen forest of central Africa, where they hunt and eat monkeys. They have little contact with other black populations, yet they have no natural immunity to the virus, and textensive tests failed to suggest that they had ever been exposed to it.

At the time of writing, no one knows where the virus came from. Some researchers believe that it is important to find out, in case there is a common reservoir for these viruses, from which more might emerge. If that reservoir could be found, it might hold clues to ways of combating the virus, perhaps with vaccines. But many researchers believe that investigations into the origin of the virus would divert valuable resources and energy away from the real problem. What is important now, they say, is to combat the epidemic and not stand around wondering where it came from.

CHAPTER 13

WASTED YEARS

Economists say that when America sneezes, Europe catches a cold. Those who study AIDS may notice a parallel. This time, however, America sneezed and nobody took any notice. While the US became riddled with pneumonia – quite literally – Europe began to feel a chill.

The story of AIDS, as far as the US is concerned, is one of missed opportunity. America has more documented cases of AIDS, by far, than any other country in the developed world. The US realized before anywhere else that it had an AIDS epidemic on its hands. It had had ample warnings of an impending catastrophe, and the technical means and money to take the necessary evasive action and educate its public. Yet the US government, unlike many European governments, has shown an astonishing lack of direction in tackling what senior health advisers have labelled the biggest public-health problem of the twentieth century.

While European governments have fought AIDS with the only effective weapon there is at present – education campaigns – the US government prevaricated and procrastinated beyond belief. As 1987 drew to a close, an estimated 1.5 million Americans were already infected with the virus. These people account for between 10 and 30 per cent of the total number of people in the world infected with HIV. By the end of 1987, nearly 50,000 Americans had developed AIDS – twice as many reported AIDS sufferers as the rest of the world put together. And yet, as governments in Europe prepared for their second and even third wave of health education campaigns, the US had barely embarked on its own nationally coordinated effort. In 1987 the US was still deliberating over plans to send education leaflets to American households.

The responsibility for this disaster must rest, to some extent, with the President, Ronald Reagan, and his administration. His shamefully inadequate actions to stop AIDS came far too late for thousands, perhaps millions, of Americans. His first public speech specifically about AIDS – before a glittering dinner at the Potomac Hotel in Washington DC in May 1987 – was littered with misguided intentions. The emphasis he chose for his fight against AIDS is extraordinary for its blind faith in technology's ability to solve the problem and its preoccupation with testing people.

On that evening of 31 May 1987, six years after scientists had first found AIDS in groups of homosexual men in California, Reagan had finally summoned up the political courage to talk about the disease. He told his audience of scientists, dignatories and film stars:

Just as most individuals don't know they carry the virus, no one knows to what extent the virus has infected our entire society. AIDS is surreptitiously spreading throughout our population, and yet we have no accurate measure of its scope. It is time we knew exactly what we were facing. And this is why I support routine testing.

I have asked the Department of Health and Human Services [HHS] to determine as soon as possible the extent to which the AIDS virus has penetrated our society and to predict its future dimensions.

I have also asked the HHS to add the AIDS virus to the list of contagious diseases for which immigrants and aliens seeking permanent residence in the United States can be denied entry.

I have asked the Department of Justice to plan for testing all Federal prisoners, as well as looking into ways to protect uninfected inmates and their families.

In addition, I've asked for a review of other Federal responsibilities, such as veterans' hospitals, to see if testing might be appropriate in these areas. This is in addition to the testing already under way in our military and foreign service.

For good measure, Reagan added that he would encourage states to 'offer' routine testing for those people seeking marriage licences. 'I would like to think,' he said, 'that everyone getting married would want to be tested.' Reagan's audience, save for a few scientists, applauded his pronouncement. His vice-president, George Bush, did not receive such a polite response a day later, when he delivered the same message at the opening of the Third International Conference on AIDS. His statements that 'there must be more testing' met with derision from many in the audience of scientists who had travelled to Washington for this annual conference, which has developed into the largest international event of its kind in just three years.

There are grave problems with routine screening for infection with HIV. To start with, there are technical problems: the test cannot be 100 per cent accurate. Another difficulty is the long period of latency between infection with the virus and the appearance of the antibodies identified by the test. In addition, and most important of all, people need to ask what will be gained by mandatory and routine screening of large groups of the population who are not at any special risk of AIDS. To a great extent, the tests become more accurate with high-risk groups, where the virus is more prevalent. Testing the population at large, therefore, for no obvious reason other than 'the government says it must be done' merely compounds the problems raised by the test.

Even before Reagan made his speech, the World Health Organization had already addressed some of these difficulties. A month previously, the WHO had published a report on routine screening of foreign visitors. The WHO identified a major problem with testing large numbers of international travellers.

Suppose, the WHO said, that in a group of 1 million travellers, there are 10,000 (1 per cent) who are carriers for the virus, in other words true HIV positives. The rest, 990,000 are true negatives. Now suppose that an immigration authority of a country began testing these 1 million visitors with a blood test that could identify true positives 99 per cent of the time (its 'sensitivity', as described in Chapter 5). The test would find 9,900 of the true positives, but fail to identify 100 carriers of the virus, who would be free to visit the country in question.

If the same test could accurately identify true negatives 99 per cent of the time (its 'specificity'), then the test would correctly label 980,100 out of 990,000 true negatives. Furthermore, it would wrongly classify 9,900 negative people as HIV positive.

So, in this hypothetical situation the test would label 19,800 people as positive. But half are true positives, and half are false positives. In other words, the test, used to conduct mass screening without confirmatory testing, is no better than flipping a coin at predicting who is really infected with the virus that causes AIDS. Furthermore, the test has failed to identify 100 true HIV positive people. And this is with a test that is 99 per cent specific and sensitive – most testing would be less accurate under the difficult conditions of quick mass screening of foreigners.

The WHO said that mass testing of foreign travellers 'can lead to massive misallocation of resources'. These resources

would be more effectively directed to educating the population concerning HIV or to screening of blood for transfusion. A less obvious negative effect could be a false sense of security about seronegative travellers, leading to laxity regarding behaviours which spread the virus and an actual increase in overall risk for HIV transmission from international travellers to national residents. Finally, it is quite possible that HIV screening of international travellers, if practised on a selective geographical basis, would provide a disincentive for the reporting of AIDS, resulting in further distortion of the critical surveillance function necessary for ongoing monitoring of the worldwide epidemic.

Routine testing of foreign visitors, in other words, can be positively harmful. By the time Reagan had made his speech, the WHO had made it abundantly clear that it opposed the sort of mandatory screening programmes that he now publicly supported: 'There are more effective, less intrusive and less costly measures for preventing HIV transmission than the use of mandatory universal screening,' the WHO said.

President Reagan, however, ignored these powerful arguments against enforced screening. The US government introduced legislation in the summer of 1987 to screen potential immigrants, although it stopped short of plans to screen tourists and foreign businessmen. The government set aside hundreds of millions of dollars for further testing programmes elsewhere in government that had nothing to do with the direct health of the people involved. Reagan's accountants estimate that he will spend $168m in 1988 on blood tests that take place outside America's health system – in the Department of Defense and Department of Labor, for instance. It is a powerful sign to private organizations they too should accept the necessity of mass screening on a routine basis.

America of course is not alone in its fascination with the test for infection with HIV. Many countries, of all political persuasions, have instigated mandatory screening programmes of one sort or another. A related issue is whether doctors have the right to test their patients for HIV infection without their patients' consent. In Britain, doctors voted at the British Medical Association's annual conference in July 1987 for secretive tests to protect themselves from possible infection. After the vote, the chairman of the BMA's council, John Marks, took legal advice. The BMA's lawyers said that testing for HIV infection without the knowledge or consent of the doctors' patients would constitute 'an invasion of the patient's bodily integrity' and is therefore grounds for the patient to sue the doctor for assault. The council quickly ruled that it would not be in the interests of the BMA to condone such surreptitious tests, and so overruled the vote of its own membership.

In the same year, the American Medical Association, the equivalent of the BMA in the US, rejected secret tests. The

AMA also passed a resolution at its annual meeting to protect the confidentiality of patients tested for HIV infection. But only, the association said, if that privacy does not infringe another person's right to safety. If a man refuses to tell his wife that he is HIV positive, then the doctor can tell her in order to protect her health.

The American Medical Association takes the view that people who are HIV positive have the right to be 'free from irrational acts of prejudice', but, equally, others have the right to protection against 'an unreasonable risk of disease'.

Testing for HIV infection is therefore a balancing act between the rights of the individual and the rights of society and others at large. Two doctors from the New England Medical Center in Boston, Klemens Meyer and Stephen Pauker, summed up the dilemma and highlighted their worries over testing in an article they wrote for the *New England Journal of Medicine* in July 1987: 'We are a testing culture: we test our urine for drugs; we test our sweat for lies. It is not surprising that we should test our blood for the acquired immune deficiency syndrome.' They warned that inaccurate tests, and tests are always inaccurate to some extent, can turn a screening programme into a social catastrophe. 'An AIDS epidemic frightens us all,' they said,

> but we should not allow our fear to cloud our judgement. Hasty and indiscriminate screening for antibody to HIV is imprudent and potentially dangerous, whether we suggest the tests to young women, require them of engaged couples, or impose them on our veterans . . . If we want to test each other, we should make a deliberate choice of the threshold probability of infection above which we will screen. We should make explicit the trade-offs in any testing programme. How many engagements should end to prevent one infection? How many jobs should be lost? How many insurance policies should be cancelled or denied? How many fetuses should be aborted and how many couples should remain childless to avert the birth of one child with AIDS?

Such are the penalties for a culture obsessed by the search for the virus.

In Africa, in contrast to the US and Europe, the testing mentality has yet to take hold. One reason is probably that testing equipment has been scarce since the AIDS epidemic began. In some countries, anyone who wanted to carry out a test for HIV antibodies had to send samples of blood back to Europe. This deficiency has not seriously hampered the diagnosis of AIDS, however. Except in unusual cases, an experienced doctor can easily recognize AIDS without performing any laboratory tests at all. Even if tests for antibodies to the virus were widely available, they might be of little use to doctors working in isolated hospitals. By 1987, as more and more sophisticated tests came on to Western markets, no one had devised a test suitable for use throughout Africa. Existing tests were either too expensive for African health budgets or too complicated to handle in the field.

Nevertheless, by 1987, many African countries were able to begin testing samples of their populations to assess how extensively the infection was spreading. By the end of that year, the World Health Organization's Special Programme on AIDS was working with virtually every country in sub-Saharan Africa. Countries taking part, including many others around the world, set up their own national committees and design their own five-year plans to combat the spread of the virus. The World Health Organization helps to arrange donations from sponsoring countries where necessary, to pay for the expense involved. In 1987, for example, the organization raised $21m to fund the first year of national programmes to stop the further spread of AIDS in Uganda, Ethiopia, Kenya, Rwanda and Tanzania.

By 1 November 1987, the World Health Organization had received requests for collaboration from 127 countries. With some countries, however, persuading governments to admit that AIDS does pose a problem within their borders has been an uphill struggle. Jonathan Mann, director of the Special Programme on AIDS, says that countries everywhere have delayed acting against AIDS. Some are still at the point where they are minimizing and denying the problem. Others are at the stage of 'reluctant acceptance'. Lastly comes 'constructive engagement', when the country starts to grapple seriously with the problem, with the launch of wide-ranging programmes to educate the public in health issues.

Constructive engagement usually comes about when the people in power realize that 'it could happen to them'. There has been much resentment that Western governments waited until AIDS was no longer a disease of minorities before acting to contain its spread. Many people, particularly in the US, believe that governments were worried only about the risk of AIDS hitting 'the general population'. While the virus stayed within groups more or less marginalized from the rest of society, the people in power could afford to ignore it. It is as though the epidemic, having encountered the revulsion of Western society's 'moral standards', has responded by ricocheting against the barriers that such prejudices foster, sending splinters everywhere. If action had been taken earlier, it might have been possible at least to limit the spread of the virus.

As it is, prejudices and taboos are everywhere paralysing the will to act. Some parents would rather ignore the fact that their children might be sexually active than tell them how to reduce their risk of catching AIDS by using a condom. Some authorities would prefer to carry on denying drug addicts clean needles in the false assumption that this will stop addicts from injecting drugs. And some prime ministers and presidents prefer to avoid using the word 'homosexual', as if, by pretending that homosexuals did not exist, they might go away.

An analogy can be drawn with a longstanding battle to repel an invading army. AIDS first attacked homosexuals offshore. That battle has already been fought. The fighting continues on the beaches, with drug addicts. There are signs that that battle will be lost. Finally, AIDS threatens to reach the land, and attack heterosexuals. This will be the hardest battle of all to win.

If governments are to stop AIDS, they must reach out to the margins of society. The only way to stop the spread of AIDS among drug addicts, for example, is to provide them with clean needles. In Scotland, over half the drug addicts in

Edinburgh are infected. Researchers believe that the figure is so high because it is the policy of the police to search addicts, confiscating needles and syringes. The result is that many addicts are forced to share needles with others, facilitating the spread of HIV. Yet plans in several British cities to give addicts clean needles in return for used ones have sometimes met with resistance. Those who oppose such schemes, however, should ask which is worse, appearing to condone drug addiction by supplying the necessary equipment, or turning a blind eye to a breeding ground for the virus?

Another breeding ground is sexually active heterosexuals. The 'moral majority' preach monogamy within marriage and chastity outside it. In the past, society could afford to be hypocritical about sex, pretending that youngsters stayed virgins until marriage while knowing that this was rarely the case. But it is no longer tenable for society to ignore premarital relationships. The threat of the virus gaining hold in young people who are experimenting with sex is too great. Health education could be the only weapon in the battle against AIDS in this vulnerable group.

Britain's health education campaign has been held up as a model to other countries. Critics have labelled it as either tasteless or too muted, depending on their standpoint. Yet its frankness and popular approach has made 'condom' a household word. In contrast, rumour has it that the word condom became known in the White House as the 'C-word', because of President Reagan's reluctance to talk about the prophylactic.

It is impossible to predict how the epidemic will develop in the future. Will there, for example, be an epidemic of AIDS in the heterosexual populations of the US and Europe? Some scientists believe that there may not be. They suggest that there may be reasons to account for the very rapid spread of the virus among heterosexuals in Africa. For example, there is evidence to suggest that relatively high levels of other sexually transmitted diseases may facilitate the spread of the virus in Africa. Some studies in Africa have found evidence of previous syphilis in one in four pregnant women tested. Syphilitic sores, by breaching the thin skin of the genitals, could make it easy for the virus to enter the body during heterosexual intercourse.

Other researchers suggest that because African heterosexuals, like American homosexuals, are exposed to so many viruses, parasites and other microorganisms, their immune systems are continually responding to new antigens. This continual activation of the immune system could make these people particularly susceptible to infection with the virus.

However, other scientists strongly believe that AIDS in the US and Europe is no longer a disease of high-risk groups. In Britain, the epidemic is three or four years behind the epidemic in the US. In parts of New York in 1987, about one in fifty men applying to join the US army had antibodies to the virus. So a woman applicant to the army, if she decided to choose a boyfriend from among the male applicants, had a one in fifty chance of drawing a short straw. If she had more than one sexual partner, her risks of encountering the virus were obviously even higher.

Most researchers agree that if no action is taken to stop the global spread of AIDS, there will be a huge disaster. Parts of Africa are already badly affected. Take the situation in Kampala, the capital of Uganda, in 1987. If a man or a woman there chooses to have five sexual partners in his or her lifetime, one of them is likely to be infected with the virus. In many cultures, Western ones included, five sexual partners in a lifetime is not unusual. Also in Kampala, studies have shown that the proportion of pregnant women with antibodies to HIV almost doubled between early 1986 and February 1987. In February 1987, just over 24 per cent of pregnant women – almost one in four – had antibodies to the virus. In Nairobi, in Kenya, about 70 per cent of prostitutes are infected with HIV.

Statistics such as these are chilling. Where urban areas are so severely affected, catastrophe seems almost inevitable.

Some researchers who have worked in Africa are unwilling to speculate about the future; they cannot bear to contemplate what might happen. Predictions of calamity, not surprisingly, are politically unpopular. Some scientists, particularly expatriates, have had to go to the lengths of publishing their work anonymously, or risk being thrown out of the country concerned.

One key article of this kind appeared in the German scientific journal *AIDS-Forschung* in January 1987. The author, who insisted on anonymity, pointed out that one of the worst aspects of the AIDS epidemic in Africa is its effect on women of child-bearing age and their children. AIDS, he said, will drastically reduce female fertility and increase the already high death rate in infants and children, perhaps to the level where the population begins to fall.

> A high mortality amongst mothers and infants will have severe consequences on family life: surviving children will suffer through lack of care; remarriage of the father will lead only to a repetition of the cycle of ill-health and death of mother and infants. A rising number of widowers and unmarried men will lead to an increased demand for female prostitution and to further fuelling of the epidemic.

Such views are not held universally, however. Some African researchers believe that AIDS does not represent the same threat to health that it does in, say, Europe. Gottlieb Monekosso, for example, the director of the World Health Organization's regional office for Africa, said in 1987 that AIDS probably ranks only tenth or lower on a list of serious tropical diseases. Malaria, measles, diarrhoeal illnesses, tuberculosis, cholera, meningitis, yellow fever and various cancers, he said, all account for more deaths and illnesses than AIDS does. But, he admitted, 'It will change.' Others warned in 1987 that if nothing is done for ten years, however, AIDS will then be the biggest health problem in Africa.

One of the greatest ironies is that, even if a 'magic bullet' against AIDS were available, many African countries would probably not be able to afford it. Even if they could, thousands of children would continue to die in Africa as they do every year – of diarrhoea, for example – because they have been denied the most basic of public-health measures, a clean water supply. Many African scientists would probably be justified in viewing Western interest in AIDS with some suspicion: why should Western countries want to help with this disease when so many other problems have gone unassisted?

Canon Paul Oestreicher, the radical theologian, once said that the worst obscenity on earth was to believe that change was impossible. Throughout the world, not just in Africa, change is now essential. And it must be rapid. There are still many places where relatively few people are infected with the virus. Uganda, one of the first African countries to put its campaign to control AIDS into action, is providing a model for other African states. Apart from its health education campaign which urges people to 'love carefully', Uganda is also screening blood donations. Tanzania, Rwanda, Ethiopia and Kenya drew up plans for similar programmes in 1987.

Donated blood is an important source of infection with HIV in many African countries, most of which, by 1987, had no facilities for screening blood for antibodies to the virus. The issue here is lack of resources. Thomas Quinn, from the National Institute of Allergy and Infectious Diseases in the US, and his colleagues have done some interesting calculations. In some countries, they said, the cost of screening donated blood would, per transfusion, cost about thirty times the annual health budget for one person. 'In the US, approximately \$60m was spent on blood-bank screening for HIV infection in 1985, a budget many times greater than the entire health budgets of many African countries.'

In some countries, such as Zaïre, other research has suggested that medical injections probably spread the virus. In Zaïre, the health budget for the whole country, with a population of 30 million people, is less than the budget of a British district hospital serving a tenth of this number. In Zaïre, needles are reused, often on as many as thirty people, until they break.

There are solutions to these problems. Donations of medical supplies and money would go a long way towards cutting down some of these risks. The question of how to influence human sexual behaviour, in Africa as in the US and Europe, is more difficult to answer. Perhaps advertising agencies should help to devise campaigns to put over the message of safe sex in Africa.

One thing is clear. No country should shirk from implementing measures to control the spread of AIDS because they hope that science will soon provide that 'magic bullet'. We already have cures for diseases such as syphilis and gonorrhoea. So why do so many people still suffer from these damaging infections? Drugs will probably never provide a solution to infection with HIV. They would have to be taken for life, because the viral genetic material would always remain in the body's cells. The cost of treatment could cripple Western health services. In Africa, where the average health budget is about $10 per person per year, this is probably only enough to pay for just one accurate diagnostic test for HIV infection. Furthermore, any drug that has to be taken for life would be likely to have side effects, perhaps serious ones.

Perhaps one day science will provide a vaccine that can protect people from HIV. But we cannot rely on it. By 1987, many researchers were privately saying that they feared it might prove impossible to develop a vaccine. They had begun to detect signs that the antibodies that neutralize the virus might also attack our body's own cells. The treatment could be as bad as the disease.

Until science has an answer to AIDS, the only barrier that the virus will respect is less than a millimetre thick and made of rubber. This thin latex line may be only way that we have of containing the sexual transmission of AIDS.

GLOSSARY

AIDS: acquired immune deficiency syndrome, the collection of illnesses resulting from infection with the human immunodeficiency virus, HIV. Acquired because it is not inherited; immune deficiency because the immune system is seriously weakened; and syndrome because AIDS is a collection of symptoms and illnesses.

antibodies: molecules produced by B-cells in response to an antigen. By binding to the antigen, antibodies make it easier for cells such as macrophages to engulf and eliminate the foreign particle.

antigens: foreign substances or particles such as proteins, bacteria or viruses, which cause the body to produce antibodies. Viral antigens are viral components such as viral proteins.

ARV: AIDS-related virus, one of the early names for HIV, coined by scientists working in California.

B-cells: also called B-lymphocytes. They are a special type of white blood cell that can produce antibodies.

cell-mediated immunity: important in eliminating microorganisms such as viruses which live inside cells where antibodies are unable to act. It is controlled mainly by the two types of T-cell.

DNA: deoxyribonucleic acid, the genetic material which is the blueprint for living organisms.

ELISA: enzyme-linked immunosorbent assay, the type of blood test commonly used to detect antibodies to HIV.

envelope protein: the envelope protein of the human immunodeficiency virus is called gp160, which splits to form two smaller molecules, gp41 and gp120. The envelope protein appears on the surface of the virus.

Factor VIII: the protein which is necessary for the formation of blood clots in humans.

FAIDS: feline acquired immune deficiency syndrome, AIDS in cats.

false negative: when a blood test incorrectly finds no antibodies to HIV when in reality they are present.

false positive: when a blood test incorrectly finds antibodies to HIV when in reality there are none.

gene: the single unit of inheritance, common to all living things including viruses.

genetic engineering: the technology of manipulating genetic material, usually with new techniques based on splicing and inserting DNA.

glycoproteins: proteins with sugar molecules on their surfaces. The envelope proteins of the human immunodeficiency virus are glycoproteins.

haemophilia: a blood-clotting disorder usually due to a deficiency of factor VIII.

HIV: human immunodeficiency virus, the virus that results in AIDS. Different types of HIV are called HIV-1, HIV-2, etc. An international committee of virologists decided on the name in 1986.

HIV-negative: if someone is said to be HIV-negative, they have no detectable antibodies to the virus.

HIV-positive: if someone is said to be HIV-positive, they have antibodies to the virus, they are infected with the virus, and they are capable of transmitting the virus to others.

HTLV: human T-cell leukaemia/lymphoma virus, which later became known as human T-cell lymphotropic virus. Different types are called HTLV-1, HTLV-2, etc. Robert Gallo named the AIDS virus 'HTLV-3' in 1984.

immunoglobulins: antibodies.

LAV: lymphadenopathy-associated virus, the name given by Luc Montagnier to describe the AIDS virus in 1983.

lentivirus: a type of retrovirus characterized by the slowness with which it attacks its host animal. HIV is a lentivirus.

lymphocytes: white blood cells known as T-cells and B-cells.

macrophages: large mobile cells capable of engulfing and destroying microorganisms. They also have an important role in 'presenting' antigens to T-cells.

monocytes: immature, unspecialized macrophages.

mutation: a change in the genetic material which often gives rise to a change in the appearance of an organism.

mutation rate: the speed at which mutations occur.

pandemic: an extensive epidemic affecting wide geographical areas.

p24 protein: one of the inner proteins of HIV. Its presence in the blood may be important as an early indicator of the future development of AIDS.

retrovirus: a type of virus that has RNA as its genetic material, rather than DNA, and which has an enzyme to make a DNA-copy of its RNA.

reverse transcriptase: the enzyme unique to HIV that controls the manufacture of viral DNA from the viral genetic material RNA.

RNA: ribonucleic acid, another type of genetic material. Living things usually make RNA-copies of DNA.

SAIDS: simian AIDS, AIDS in monkeys.

sensitivity: the probability of a blood test giving a positive result when antibodies to HIV are in fact present. Highly sensitive tests give low false negatives.

seroconversion: the name given to the point at which someone in whom it was not previously possible to detect antibodies first produces antibodies against a microorganism.

seronegative: the status of someone who has no detectable antibodies to a particular microorganism, such as HIV.

seropositive: the status of someone who has antibodies that recognize a particular microorganism, such as HIV.

serum: the fluid that remains after the cells have been removed from the blood. It is this fluid that doctors test when they are looking for antibodies to HIV.

SIV: simian immunodeficiency virus, the virus that infects monkeys and which can cause AIDS-like symptoms in some species.

specificity the probability of a blood test giving a negative result when antibodies to HIV are in fact not present. Highly specific tests give low false positives.

T-cells: a type of white blood cell or lymphocyte. There are two main kinds of T-cell, helper cells and cytotoxic cells. The helper cells are stimulated by antigen on the surface of macrophages, and can then cause B-cells to produce antibodies, macrophages to engulf microorganisms and cytotoxic cells to destroy infected cells.

virus: one of the simplest organisms. They lead parasitic existences by hijacking the machinery of living cells.

Western blot: a type of test which is complex to perform but which is highly accurate and so is often used to confirm positive results from other, simpler tests.

INDEX